THE HEAVENS PROCLAIM

ASTRONOMY AND THE VATICAN

THE HEAVENS PROCLAIM
ASTRONOMY AND THE VATICAN

A BOOK PREPARED
BY THE MEMBERS OF THE VATICAN OBSERVATORY
ON THE OCCASION OF THE INTERNATIONAL YEAR OF ASTRONOMY

EDITED BY Br. Guy CONSOLMAGNO, S.J.
Vatican Observatory

Vatican City State
2009

VATICAN OBSERVATORY

ISBN: 978-1-59276-645-1 (Inventory No. T927)
LCCN: 2009926276

Cover and Interior Design: GianCarlo Olcuire

Cover: NGC 5128 imaged by Fr. José Funes, S.J., at Cerro Tololo Inter-American Observatory in Chile and processed by Sanae Akiyama. The stellar body of the galaxy is shown in blue and emission of the ionized gas in red.

Title Page: One of the more famous "bodies" in the sky is the Horsehead Nebula. It is a dark cloud that shadows the light of an emission nebula behind it. The "unicorn's horn" in this image is due to a bright star just out of the field of view. Image taken by Brucker, Romanishin, Tegler, and Consolmagno at the VATT.

Back Cover: The dome of the Visual Telescope at Castel Gandolfo in the moonlight. Photo by Ron Dantowitz, Clay Center Observatory.

PRINTED IN CANADA

The heavens proclaim His righteousness;
and all the peoples behold His glory.

Psalm 97:6

Above: The Milky Way setting behind the dome
of the Vatican Advanced Technology Telescope on
Mt. Graham, Arizona, on September 20, 2007. The
Moon had set just behind the VATT dome but still
lit the sky; the red glow on the building is from a
faint red light outside the entrance of the neighboring
Sub Millimeter Telescope. Photo by Jim Scotti, who
used a Canon 20D D-SLR with an 8mm focal length
lens at f/3.5; 120 second exposure at ISO 400.

CONTENTS

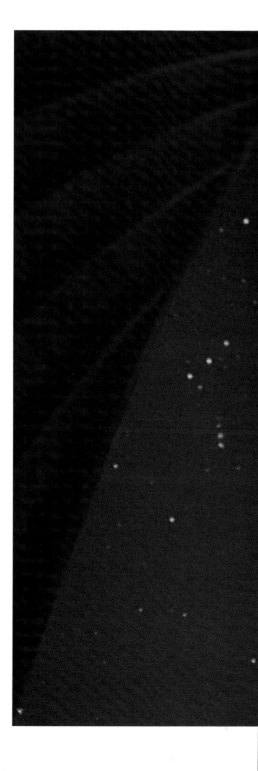

Photo credit: *Alfredo Matacotta*

INTRODUCTION

• GIOVANNI Card. LAJOLO •

Why does the starlit sky hold such a profound fascination for us? Perhaps because it is there that we encounter, commingled, the mystery of light and darkness — two primal experiences connected with the beginning and end of human life. Perhaps it comes from seeing the order, both overt and occult, in the movements of the celestial spheres, with which we sense ourselves secretly involved. Perhaps it is because we feel so small before the starry universe: like a straw tossed into the "great sea of being," we feel ourselves confronted with Destiny, and in this way we begin to become aware within ourselves, even if only in a rudimentary way, of the grand questions regarding our existence and our passing through life. Or is it perhaps for that "Love that moves the sun and the other stars," as Dante puts it?

It is our hope that through this book the reader becomes aware of just some of the depths of the universe. For millennia this has been a human quest, but only in the last few centuries has science found the proper methodological key and the tools it needs to advance with confidence.

And if the mystery of the universe has to some degree become clearer, still it remains ever more impenetrable. Because, as always, the more our understanding advances, the more its horizon toward further knowledge recedes. An unlimited vastness, no less so than that of the universe, is what human intelligence is called upon to traverse!

This book should help usher the reader into the great perspectives of present-day astronomical research. It has, however, some other particular messages that it intends to convey.

The first message, and the more important one, is to make clear that science — and in particular the science of astronomy — is not only not contrary to the revealed knowledge of faith, but it is open to it and even in some ways contiguous to it (even if to it, as we know well, is not homogeneous; it would be a great error to think that God could be encompassed and identified with objects that are appropriately studied by the means and methods of non-metaphysical sciences). To this end, besides the chapters dedicated to the knowledge of the actual universe of the stars, there is also one dedicated to the biblical vision of the cosmos, a vision not based on science but on common experience, illuminated by a poetic sentiment and by faith in the existence and the providence of God.

Another chapter is dedicated to the rapport between scientific cosmology and the topic of creation. And yet another relates significant passages drawn from the discourses of recent Roman pontiffs, from Leo XIII to Benedict XVI.

The second message, more modest but certainly no less interesting, is addressed to the world of general culture and particularly to the more restricted circle of those interested in the problems of modern astronomy who might be curious about the scientists running the Vatican Observatory and their activities. These are already well known to other specialists through their publications (listed in their *Annual Report*) and the *Vatican Observatory Newsletter*, which comes out several times a year.

The scientists of the Vatican Observatory have inherited, and are carrying forward today, a great tradition of astronomy going back to the renowned observatory of the Roman College. But they can trace their lineage back even further, to the great tradition of astronomical studies and the thousand-year interest of the Church in the subject, as is also illustrated in this book. After the 1870 occupation of Rome and the loss of the original observatory, in 1891 the great Pope Leo XIII gave new life to the observatory under the name Specola Vaticana (Vatican Observatory), which operated in the Tower of the Winds in the Vatican during the first decades of the 20th century. Then, after the reconciliation with Italy, in the years 1932-1935 Pius XI founded the existing observatory on the Pontifical Palace itself at Castel Gandolfo. An important new development occurred in the 1980s when a collaboration with the University of Arizona allowed the observatory to operate a new Vatican Advanced Technology Telescope (VATT) on Mt. Graham, at an altitude of 3,200 meters.

To the scientists of the Vatican Observatory—mostly priests of the Society of Jesus — go the greatest admiration and warm thanks on the part of the Holy See, the Governorship of Vatican City State, and me personally. I have no doubt that the readers of this book will feel likewise.

It is my pleasure to relate to them the words that, in the Gospel according to Luke, Jesus says to the seventy disciples: "…your names are written in heaven" (10:20).

And that is a "heaven" far higher than that marvelous firmament with which our scientists are so familiar — though not so familiar that they can feel themselves satisfied with what they are familiar with; because, as the book of Sirach says,

"Many things greater than these lie hidden, for I have seen but few of his works" (43:32).

*   *   *

This book is dedicated to the Holy Father, Benedict XVI, in whose residence at Castel Gandolfo is situated the headquarters of the Vatican Observatory. Some selections from his teachings on the rapport between the intelligibility of the material world, the structure of the universe, and our knowledge of God, as concerns the relationship between science and faith, make up part of a chapter in this book.

Giovanni Cardinal Lajolo
President of the Governor's Office
for Vatican City State

9

TELLING OUR STORY

• Fr. JOSÉ GABRIEL FUNES, S.J. •
Director of the Vatican Observatory

"... that everyone might see clearly that the Church and her Pastors are not opposed to true and solid science, whether human or divine, but that they embrace it, encourage it, and promote it with the fullest possible devotion."

Pope Leo XIII
The Refounding and Restructuring of the Vatican Observatory,
March 14, 1891

Sometimes you need to tell a story to explain how things have come to be ...

In a letter addressed to me, dated May 8, 2007, Cardinal Giovanni Lajolo, president of the Governatorate, invited the staff members of the Specola Vaticana (Vatican Observatory) to produce a book targeted to the general public. The cardinal's invitation was perfectly in line with the two-fold mission statement of the observatory. We have to do good science; but we must also be seen doing so — we must let the world see that the Church is supporting us as we're doing our science. This book addresses the second part of our mission.

The story that explains our existence is the Story of Creation. The stars tell us that story.

As so, in the opening chapter we offer selections from Sacred Scripture alongside astronomical images to remind us of the beauty of the universe and its Creator.

The subtitle makes the connection between astronomy and the Vatican, and expresses the will of the modern Popes to have a research institute in astronomy. Their desire is seen here in how they have addressed topics related to the study of astronomy.

Later in this book you'll find a section of Frequently Asked Questions. "Why is the Vatican interested in astronomy?" is the most frequently asked! To understand why, it is essential to tell our story. Thus, history of astronomy is an important part of our research. Three chapters here are dedicated to the reform of the calendar, to Galileo, and to the history of the Vatican Observatory.

Though we are a small staff, we cover the main fields of astrophysics: from our backyard, the Solar System, to distant galaxies and the universe itself. Some of our research in these areas is outlined here. But our present scientific knowledge and our present

ignorance of the origin of the universe pose many questions that go beyond the science. How is this beautiful universe related to the creation *ex nihilo*? A chapter here addresses this question.

Another very frequent question we are asked by visitors and journalists concerns the relationship between science and faith. The reader may wonder why there is not an explicit chapter on Astronomy and Faith. But the very fact that we are, at the same time, religious men and scientists tells more than words that science and faith are compatible. Our mere existence is witness that science and religion can live in harmony. For too many years there has been a "war" between the two sides. Dialogue, though difficult, is necessary. Is there any honest dialogue that is not difficult? There may be conflicts and tensions, but we must not be afraid of them. The Church does not fear science and its discoveries.

Pope John XXIII once said that our mission must be to explain the Church to astronomers, and astronomy to the Church. And Pope Benedict XVI, addressing the most recent General Congregation of the Jesuits, insisted that we must be "on the frontier." I believe that the Vatican Observatory has this mission: to stand on the frontier between the world of science and the world of faith, to bear witness to how it is possible to believe in God, and at the same time to be good scientists. The observatory is a bridge, a small bridge, between the Church and astronomy.

This book is being published in 2009, the International Year of Astronomy, when we celebrate the 400th anniversary of the first use of an astronomical telescope by Galileo Galilei. This ordinary-extraordinary event in human history indeed opened a new window to the exploration of the universe for everyone. Astronomy has had a huge impact on our culture. Astronomy helps us to see the beauty of the universe and to appreciate the fragility of our existence. For people of faith, it opens minds and hearts to the Creator. The Church has always understood the importance of astronomy in human culture, and has embraced, encouraged, and promoted it. The Vatican Observatory today is a concrete sign of that commitment.

We are deeply grateful to Cardinal Lajolo for supporting our work, and for encouraging us to communicate what we do. I would also like to thank Br. Guy Consolmagno, S.J., the editor, who has put so much effort and hard work into this book. May you, the reader of this book, find spiritual consolation in the contemplation of the stars that *shine to delight their Creator* (see Baruch 3:34). ●

Pontifical Palace,
Castel Gandolfo,
October 2008

11

STARS
IN SCRIPTURE

Above: *Annular solar eclipse, photographed by Sassone Corsi with the Vatican Observatory's Coronado telescope in Tunisia, October 2005.*

THE BODIES IN THE SKY ARE GOD'S CREATION.

G od made the two great lights — the greater light to rule the day and the lesser light to rule the night — and the stars. God set them in the dome of the sky to give light upon the earth, to rule over the day and over the night, and to separate the light from the darkness. And God saw that it was good. And there was evening and there was morning, the fourth day.

Genesis 1:16-19

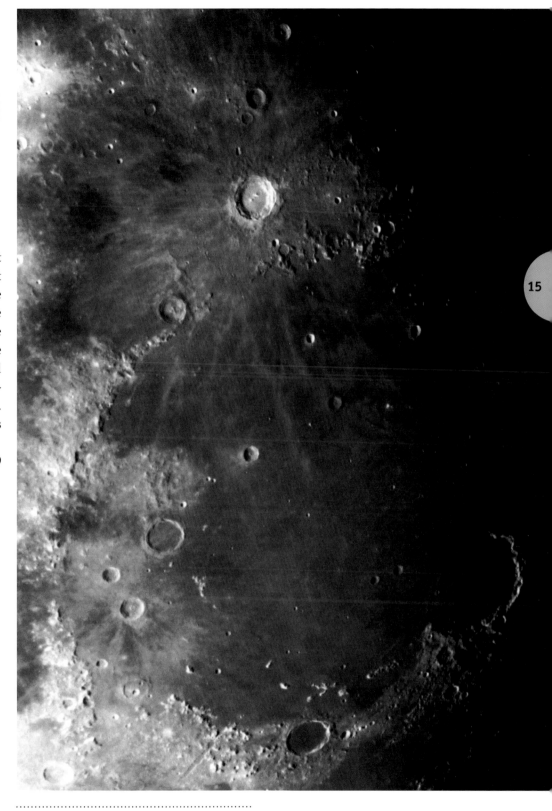

Above: Crater Copernicus, *photographed with the 40 cm Visual telescope at Castel Gandolfo by Fr. Manuel Carreira, S.J., during the 1995 Vatican Observatory Summer School.*

ALL THE
UNIVERSE
IS UNDER
GOD'S RULE.

Then Job answered: "Indeed I know that this is so; but how can a mortal be just before God? If one wished to contend with Him, one could not answer Him once in a thousand.

"He is wise in heart, and mighty in strength; who has resisted Him, and succeeded? He, Who removes mountains, and they do not know it, when He overturns them in His anger; Who shakes the earth out of its place, and its pillars tremble; Who commands the sun, and it does not rise; Who seals up the stars; Who alone stretched out the heavens and trampled the waves of the Sea; Who made the Bear and Orion, the Pleiades and the chambers of the south; Who does great things beyond understanding, and marvelous things without number."

Job 9:1-10

Top: *The constellation of Orion — note the telltale three stars of Orion's belt — is seen in this time exposure setting behind the dome of the Vatican Advanced Technology Telescope on Mt. Graham, Arizona, USA.*
Photo credit: *Roelof de Jong, now at the Space Telescope Science Institute.*

Opposite: *A thin section of a meteorite that fell from the asteroid belt and landed on Earth near the town of Knyahinya, Ukraine, in 1866; imaged through transmitted polarized light by Br. Guy Consolmagno using the optical microscope in the Vatican Observatory's meteorite laboratory in Castel Gandolfo.*

P raise the LORD!
How good it is to sing praises to our God;
for He is gracious, and a song of praise is fitting.

The LORD builds up Jerusalem;
He gathers the outcasts of Israel.

He heals the brokenhearted,
and binds up their wounds.

He determines the number of the stars;
He gives to all of them their names.

Psalm 147:1-4

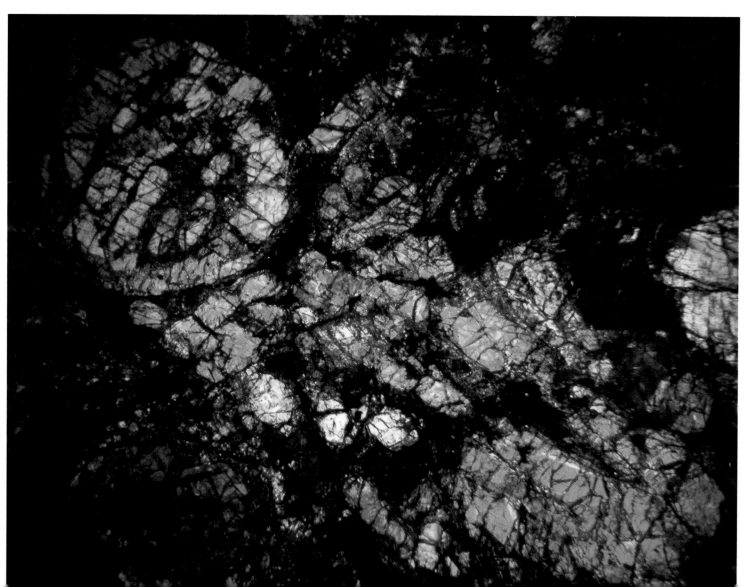

Then the LORD answered Job out of the whirlwind: "Who is this that darkens counsel by words without knowledge? ... I will question you, and you shall declare to me. Where were you when I laid the foundation of the earth? Tell me, if you have understanding. Who determined its measurements — surely you know! Or who stretched the line upon it? On what were its bases sunk, or who laid its cornerstone, when the morning stars sang together and all the heavenly beings shouted for joy?"

Job 38:1-7

STARS ALSO
PRAISE
AND ADORE
GOD.

Praise the LORD!
Praise the LORD from the heavens;
praise Him in the heights!

Praise Him, all his angels;
praise Him, all His host!

Praise Him, sun and moon;
praise Him, all you shining stars!

Praise Him, you highest heavens,
and you waters above the heavens!

Let them praise the name of the LORD,
for He commanded and they were created.

He established them forever and ever;
He fixed their bounds, which cannot be passed.

Psalm 148:1-6

O Israel, how great is the house of God, how vast the territory that He possesses! It is great and has no bounds; it is high and immeasurable....

Who has gone up into heaven, and taken her, and brought her down from the clouds?

... the stars shone in their watches, and were glad; He called them, and they said, "Here we are!" They shone with gladness for Him who made them.

This is our God; no other can be compared to Him.

Baruch 3:24-25, 29, 34-35

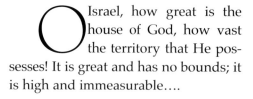

Above: The Abell 397 cluster of galaxies, imaged
at the Vatican Advanced Technology Telescope
by Matt Nelson in 2001, combining a set of four
V (green) filter, five R (red) filter, and seven B (blue)
filter images taken over the course of about an hour.
The history of each filter is recorded in the trail
of an asteroid moving right to left from the center
of the picture: seen at first through green,
then red, then blue filters.

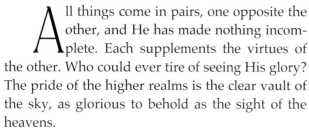

STARS ARE EVIDENCE OF GOD'S GLORY.

All things come in pairs, one opposite the other, and He has made nothing incomplete. Each supplements the virtues of the other. Who could ever tire of seeing His glory? The pride of the higher realms is the clear vault of the sky, as glorious to behold as the sight of the heavens.

The sun, when it appears, proclaims as it rises what a marvelous instrument it is, the work of the Most High. At noon it parches the land, and who can withstand its burning heat? A man tending a furnace works in burning heat, but three times as hot is the sun scorching the mountains; it breathes out fiery vapors, and its bright rays blind the eyes. Great is the Lord who made it; at His orders it hurries on its course.

It is the moon that marks the changing seasons, governing the times, their everlasting sign. From the moon comes the sign for festal days, a light that wanes when it completes its course. The new moon, as its name suggests, renews itself; how marvelous it is in this change, a beacon to the hosts on high, shining in the vault of the heavens! The glory of the stars is the beauty of heaven, a glittering array in the heights of the Lord. On the orders of the Holy One they stand in their appointed places; they never relax in their watches.

Sirach 42:24–43:10

O LORD, our Sovereign,
    how majestic is Your name in all the earth!
    You have set Your glory above the heavens....

When I look at Your heavens, the work of Your fingers,
the moon and the stars that You have established;

what are human beings that You are mindful of them,
mortals that You care for them?

Yet You have made them a little lower than God,
and crowned them with glory and honor.

Psalm 8:1, 3-5

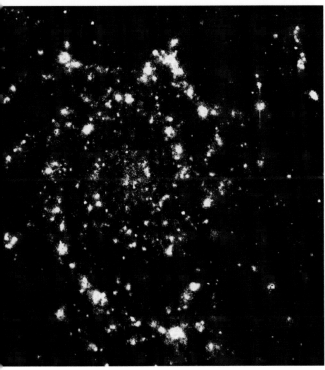

*"All things
come in pairs."
Sometimes a pair of images
can help us understand
in a way we never could
with just one image.*

Above: *The globular cluster M2 as imaged at the VATT by Melissa Brucker and Bill Romanishin
of the University of Oklahoma, Steve Tegler of Northern Arizona University, and Br. Guy Consolmagno,
S.J., of the Vatican Observatory. Advanced CCD cameras can capture a remarkable dynamic range of light,
from faint to bright; here the same image is reproduced at two different contrasts to show the faint outer
stars (left) and the structure in the core (right) of this cluster.*
Below: *The spiral galaxy NGC 628 in red (left) and H-alpha (right) filters, imaged by Fr. José Funes, S.J.
The H-alpha spots shine where new stars are being formed.*

GOD GIVES US WISDOM TO UNDERSTAND HIS STARS.

May God grant me to speak with judgment, and to have thoughts worthy of what I have received; for He is the guide even of wisdom and the corrector of the wise. For both we and our words are in His hand, as are all understanding and skill in crafts.

For it is He who gave me unerring knowledge of what exists, to know the structure of the world and the activity of the elements; the beginning and end and middle of times, the alternations of the solstices and the changes of the seasons, the cycles of the year and the constellations of the stars, the natures of animals and the tempers of wild animals, the powers of spirits and the thoughts of human beings, the varieties of plants and the virtues of roots; I learned both what is secret and what is manifest, for wisdom, the fashioner of all things, taught me.

There is in her a spirit that is intelligent, holy, unique, manifold, subtle, mobile, clear, unpolluted, distinct, invulnerable, loving the good, keen, irresistible, beneficent, humane, steadfast, sure, free from anxiety, all-pow-

Right: This composite by Br. Karl Treusch, S.J., is taken from The Green Flash and Other Low Sun Phenomena *by Fr. Daniel O'Connell, S.J. This 1958 book presented the first published photographs of the "Green Flash," confirming that it was a real phenomenon, not an optical illusion. Observations of the Green Flash have led to new insights into atmospheric turbulence and structure.*

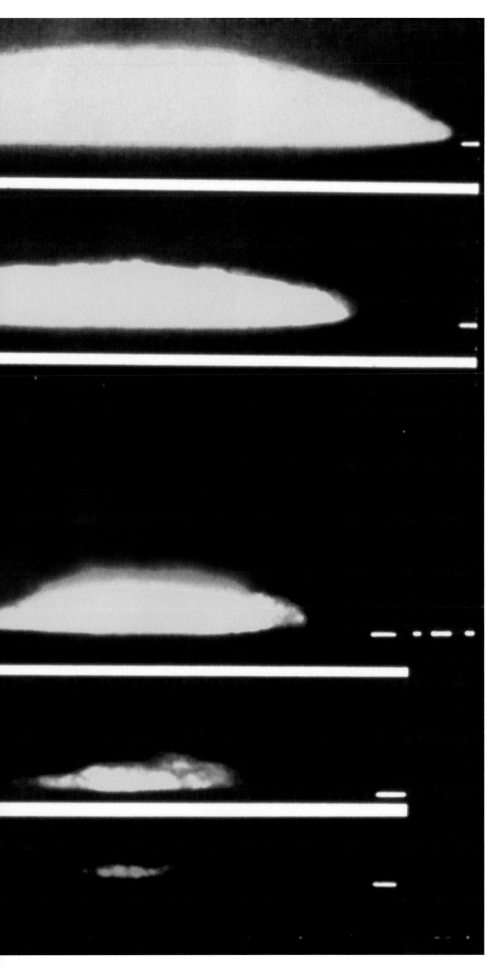

erful, overseeing all, and penetrating through all spirits that are intelligent, pure, and altogether subtle. For wisdom is more mobile than any motion; because of her pureness she pervades and penetrates all things.

For she is a breath of the power of God, and a pure emanation of the glory of the Almighty; therefore nothing defiled gains entrance into her. For she is a reflection of eternal light, a spotless mirror of the working of God, and an image of His goodness. Although she is but one, she can do all things, and while remaining in herself, she renews all things; in every generation she passes into holy souls and makes them friends of God, and prophets; for God loves nothing so much as the person who lives with wisdom.

She is more beautiful than the sun, and excels every constellation of the stars. Compared with the light she is found to be superior, for it is succeeded by the night, but against wisdom evil does not prevail.

Wisdom 7:15-30

STARS ARE TO BE APPRECIATED, BUT NOT TO BE WORSHIPED.

For all people who were ignorant of God were foolish by nature; and they were unable from the good things that are seen to know the One Who exists, nor did they recognize the artisan while paying heed to His works; but they supposed that either fire or wind or swift air, or the circle of the stars, or turbulent water, or the luminaries of heaven were the gods that rule the world. If through delight in the beauty of these things people assumed them to be gods, let them know how much better than these is their Lord, for the author of beauty created them. And if people were amazed at their power and working, let them perceive from them how much more powerful is the One Who formed them. For from the greatness and beauty of created things comes a corresponding perception of their Creator.

Wisdom 13:1-5

And when you look up to the heavens and see the sun, the moon, and the stars, all the host of heaven, do not be led astray and bow down to them and serve them, things that the LORD your God has allotted to all the peoples everywhere under heaven.

Deuteronomy 4:19

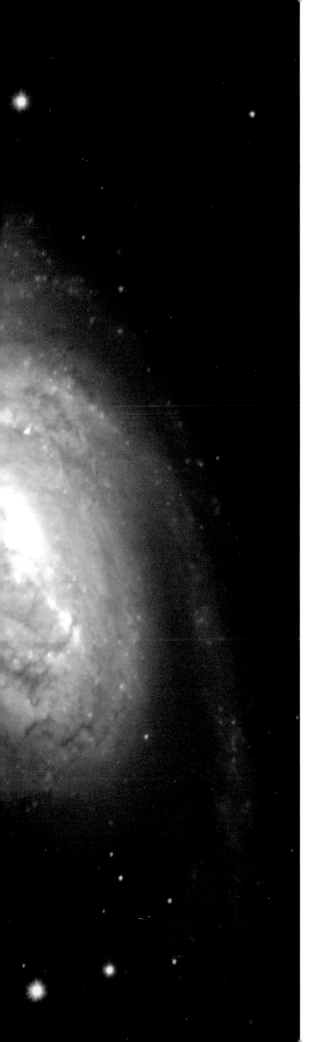

Left: NGC 2903, a galaxy of about a billion stars
some 80,000 light-years in diameter, is located
20 million light-years from us. Note the faint blue
arms extending out from each side of the spiral's
center; the central regions have an unusual
amount of star formation. This image was made
at the VATT by Brucker, Romanishin, Tegler,
and Consolmagno, combining exposures
taken through blue, visual (green), and red filters.

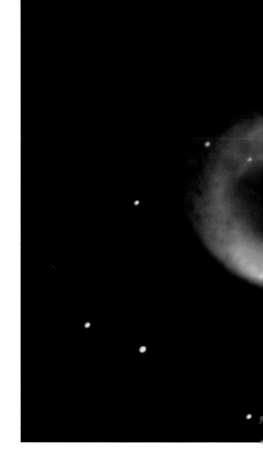

STARS ARE SIGNS OF GOD'S ETERNAL LOVE.

O give thanks to the LORD, for He is good,
for His steadfast love endures forever.

O give thanks to the God of gods,
for His steadfast love endures forever.

O give thanks to the Lord of lords,
for His steadfast love endures forever;

Who alone does great wonders,
for His steadfast love endures forever;

Who by understanding made the heavens,
for His steadfast love endures forever;

Who spread out the earth on the waters,
for His steadfast love endures forever;

Who made the great lights,
for His steadfast love endures forever;

The sun to rule over the day,
for His steadfast love endures forever;

The moon and stars to rule over the night,
for His steadfast love endures forever....

Psalm 136:1-9

Circles are an ancient symbol of eternity.

The rings here are two planetary nebulae.
Top: *The Ring Nebula of Lyra, M57, 2,300 light-years away, as imaged by Matt Nelson of the University of Virginia.*
Opposite: *The Helix Nebula, NGC 7293, in Aquarius, which is 700 light-years from us; imaged by Brucker, Romanishin, Tegler, and Consolmagno. Both images were taken at the Vatican Advanced Technology Telescope on Mt. Graham, Arizona.*

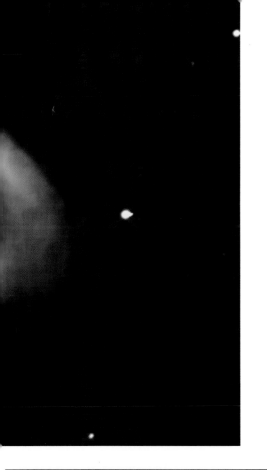

A WORLD
WITHOUT
STARS IS
A TRAGEDY...

See, the day of the LORD comes, cruel, with wrath and fierce anger, to make the earth a desolation, and to destroy its sinners from it. For the stars of the heavens and their constellations will not give their light; the sun will be dark at its rising, and the moon will not shed its light. I will punish the world for its evil, and the wicked for their iniquity; I will put an end to the pride of the arrogant, and lay low the insolence of tyrants.

Isaiah 13:9-11

In the twelfth year, in the twelfth month, on the first day of the month, the word of the LORD came to me:

Mortal, raise a lamentation over Pharaoh king of Egypt, and say to him: You consider yourself a lion among the nations, but you are like a dragon in the seas; you thrash about in your streams, trouble the water with your feet, and foul your streams. Thus says the Lord GOD: In an assembly of many peoples I will throw my net over you; and I will haul you up in my dragnet....

When I blot you out, I will cover the heavens, and make their stars dark; I will cover the sun with a cloud, and the moon shall not give its light. All the shining lights of the heavens I will darken above you, and put darkness on your land, says the Lord GOD. I will trouble the hearts of many peoples, as I carry you captive among the nations, into countries you have not known.

Ezekiel 32:1-3, 7-9

Opposite: When stars consume all of their hydrogen fuel, they can collapse and explode in dramatic ways. Smaller stars erupt disks of gas called Planetary Nebulae: at right are (above) the Blinking Nebula (NGC 6826); (middle) the Saturn Nebula (NGC 7009); and (bottom) the Dumbbell Nebula. Top and middle images are by Brucker, Romanishin, Tegler, and Consolmagno; bottom image by Matt Nelson. All images taken at the VATT.

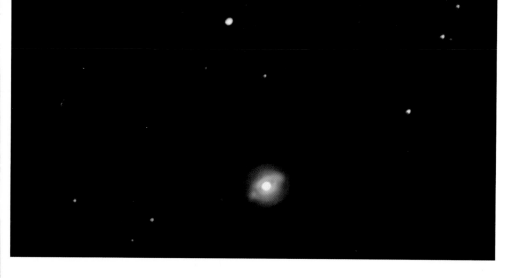

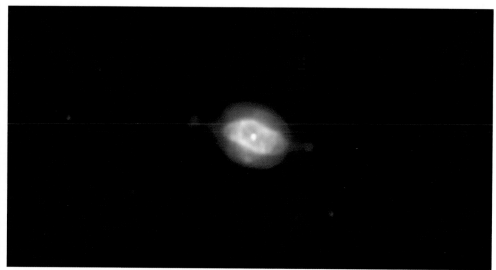

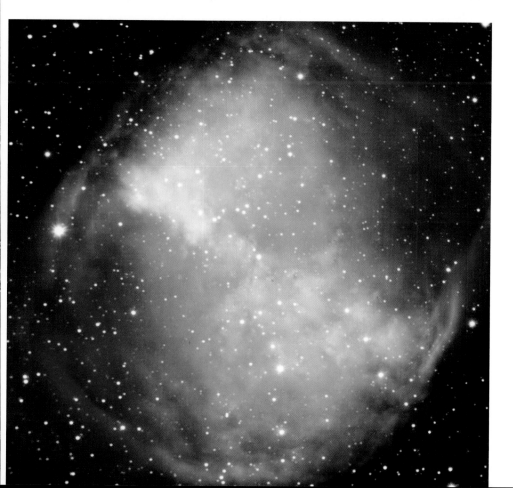

Above: *A massive star can erupt into a supernova like the Crab Nebula, imaged at the VATT by Fr. Richard Boyle, S.J.*

... A WORLD
WITHOUT
STARS IS
A TRAGEDY.

The earth quakes before them, the heavens tremble. The sun and the moon are darkened, and the stars withdraw their shining. The LORD utters His voice at the head of His army; how vast is His host! Numberless are those who obey His command. Truly the day of the LORD is great; terrible indeed — who can endure it? ... Multitudes, multitudes, in the valley of decision! For the day of the LORD is near in the valley of decision. The sun and the moon are darkened, and the stars withdraw their shining. The LORD roars from Zion, and utters His voice from Jerusalem, and the heavens and the earth shake. But the LORD is a refuge for His people, a stronghold for the people of Israel.

Joel 2:10-11; 3:14-16

Immediately after the suffering of those days the sun will be darkened, and the moon will not give its light; the stars will fall from heaven, and the powers of heaven will be shaken.

Matthew 24:29

There will be signs in the sun, the moon, and the stars, and on the earth distress among nations confused by the roaring of the sea and the waves.

Luke 21:25

But in those days, after that suffering, the sun will be darkened, and the moon will not give its light, and the stars will be falling from heaven, and the powers in the heavens will be shaken. Then they will see "the Son of Man coming in clouds" with great power and glory.

Mark 13:24-26

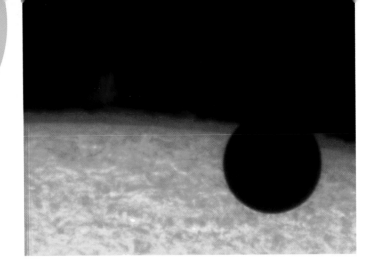

B
ut God gives it a body as He has chosen, and to each kind of seed its own body. Not all flesh is alike, but there is one flesh for human beings, another for animals, another for birds, and another for fish. There are both heavenly bodies and earthly bodies, but the glory of the heavenly is one thing, and that of the earthly is another. There is one glory of the sun, and another glory of the moon, and another glory of the stars; indeed, star differs from star in glory. So it is with the resurrection of the dead. What is sown is perishable, what is raised is imperishable.

1 Corinthians 15:38-42

Within our Solar System are two very different types of bodies: stars, and planets. And one planet in particular is special to us: our home planet, Earth.
Above: *Sunrise over Lake Albano, Castel Gandolfo.*
Top: *The planet Venus beginning its transit across the face of the Sun on June 8, 2004. Both photos were taken outside the dome of the Double Astrograph telescope in Castel Gandolfo by Ron Dantowitz, Clay Center Observatory.*

STARS IN OUR HISTORY

Above: *Comet Whipple-Fedtke-Tevadze (C/1942 X1), scanned from the glass plate taken as a 40-minute exposure at the Double Astrograph refractor camera in Castel Gandolfo by Fr. Josef Junkes, S.J., and Fr. Peter Zirwes, S.J., on the evening of January 27, 1943.*

ASTRONOMY, CALENDARS, AND RELIGION

• Fr. JUAN CASANOVAS, S.J. •

GREGORIANVM. 53

OCTOBER.

Cui defunt decem dies pro correctione Anni Solaris.

Cyclus E-pact. Anni correction. MDLXXXII	Lit. Do-mi-nic.		Dies Men fis.	
xxij	A	Kal.	1	Remigii Episcopi & Confeff.
xxj	b	vi	2	
xx	c	v	3	
xix	d	iiij. No.	4	*Francisci Confeff.* duplex.
viij	A	Idib.	15	Dionyfii, Ruftici, & Eleutherii martyrum. femiduplex. cum commemoratione S. Marci Papæ & Confefto-ris, & SS. Sergii, Bacchi, & Apuleii martyrum.
vij	b	xvii	16	Calixti Papæ, & mart. femiduplex.
vj	c	xvj	17	
v	d	xv	18	*Lucæ Euangeliftæ.* dupl.
iiij	e	xiiij	19	
iij	f	xiij	20	
ij	g	xij	21	Hilarionis Abbatis. & comm. SS. Vrfulæ & fociarum virg. & mart.
j	A	xj	22	
✳	b	x	23	
xxix	c	ix	24	
xxviiij	d	viij	25	Chryfanti, & Dariæ marty.
xxvij	e	vij	26	Euarifti Papæ & mart,
xxvi	f	vj	27	Vigilia
25 xxv	g	v	28	*Simonis & Iudæ Apoftolorum.* dup.
xxiiij	A	iiij	29	
xxiij	b	iij	30	
xxii	c	Pr. Kal.	31	Vigilia

The Hebrew Calendar

The first and most practical reason why people in antiquity studied the stars and planets, and their relative motions, was to construct a reliable calendar.

A calendar counts days in units of weeks, months, and years. The challenge of making any calendar comes from the fact that the lunar month does not have an exact number of days; and likewise, the year has neither an exact number of days nor an exact number of months. The differences, fractions of a day and fractions of a month within a year, accumulate over the years, and it eventually becomes necessary to add or subtract days in order to keep the calendar in harmony with the annual sequence of the seasons. This operation is called *interpolation*. A good calendar is a set of simple and clear rules in order to know when to interpolate a day or even a lunar month when needed.

The familiar seven-day week probably originated in Babylon, predating written history; one theory suggests it was a way of counting the days

Left: The calendar for the month of October 1582 from Christoph Clavius' book explaining the reformed Gregorian calendar. In order to bring the calendar in line with the seasons, the days between October 4 and October 15 were skipped that year.

between market days. The month was originally based on the phases of the Moon; knowing when the Moon would be full (and thus light the nighttime sky) was especially important for hunters or anyone living near tidal waters (the highest tides being associated with New and Full Moons). The year was most important in agricultural societies, where knowledge of planting and harvesting times was needed.

For the ancient Hebrews, the calendar was not merely a table of numbers put together by human beings, but it was founded on celestial phenomena whose origins were deemed to come from God. You can see this in the way the Bible describes the ultimate purpose of the Creator in making celestial objects: in Genesis (1:14-18) we are told that "God said, 'Let there be lights in the dome of the sky to separate the day from the night; and let them be for signs and for seasons and for days and years, and let them be lights in the dome of the sky to give light upon the earth.' And it was so. God made the two great lights — the greater light to rule the day and the lesser light to rule the night — and the stars. God set them in the dome of the sky to give light upon the earth, to rule over the day and over the night, and to separate the light from the darkness." Likewise, we find in Psalm 104:19, "You have made the moon to mark the seasons." Thus the original Hebrew calendar was controlled by celestial phenomena, phenomena that everyone could easily see.

The actual length of a lunar month (the time over which the Moon cycles through all of its phases) is twenty-nine and a half days (29.53085 days according to the presently accepted value). The Hebrews defined their year to consist of twelve months of alternating 29 and 30

ÉQUINOXE DE MARS

Sens du mouvement de Translation

PRINTEMPS

Avril

Mars

Mai

Février

Juin

Janvier

Périhélie

SOLSTICE DE JUIN

Ligne des

Ligne des Apsides

Aphélie

SOLSTICE DE DÉCEMBRE

HIVER

Juillet

Décembre

Août

9bre

AUTOMNE

7bre

8bre

Cercle parfait montrant l'ellipticité de l'orbite

ÉQUINOXE DE SEPTEMBRE

days' length. The problem is that such a *lunar year* has only 354 days, and so it is about 11 days too short compared to a *solar year* that follows the seasons. But farmers needed to keep track of the sea-

....

Above: The annual movement of the Earth around the Sun and the production of the seasons, as illustrated in C. Flammarion's book Astronomie Populaire, *published in Paris in 1880.*

sons, to regulate agricultural activities such as sowing and harvesting crops. In any calendar designed to follow both the Moon and the Sun (a *lunisolar* calendar), therefore, it is necessary to introduce an extra, *intercalary* month every two or three years.

In 432 B.C., the Greek mathematician Meton noticed that 19 solar years adds up to almost exactly 235 lunar months; the difference is only a couple of hours, slipping by only about a day over a span of about 300 years. Thus one can coordinate the lunar and solar calendars by inserting seven intercalary (or *embolistic*) months over this 19-year pattern. It is easy enough to construct a table for each year of the Metonic cycle, giving the day in which there is a New Moon. As the cycle repeats, you would have an almost perpetual calendar. Meton devised such a rule for when to insert these extra months.

But the ancient Hebrews were not aware of the Metonic system. In the absence of a precise rule, knowing when to insert an extra intercalary month was not trivial. An expert in astronomy could easily ascertain the position of the Sun with a simple gnomon (you can observe the changing length of the shadow of the Sun as projected by a vertical pole or an obelisk), but it seems that in practice the Sanhedrin, the assembly of elders who regulated Jewish life, were not particularly systematic in their working out the progress of the farmers' seasons. Thus there was a certain uncertainty as to when they would declare the first Moon (or month) of the year; intercalary months were inserted whenever it was deemed opportune. They found it easier to keep track of the years according to particular events that occurred within a year, rather than count from a precisely determined beginning of the year.

The year always began on the first day of a lunar month. Usually, the first month of the year was set to coincide with springtime (in some places it was set to begin with autumn, but it's the same idea as far as this discussion is concerned). The lunar month always began with the New Moon. The day of the New Moon, the *Neomenia*, was a holiday with certain prescribed sacrifices at the Temple (cf. Numbers 28:11-15).

Finding the beginning of the month was not always easy. Aside from the problem of bad weather hiding the Moon, it is impossible to actually observe a New Moon itself since it occurs when that body is in conjunction with the Sun, and so is hidden by the Sun's brilliance. You can only see the thin crescent of the Moon on the day after it is new, and even that is not easy. Even today, Muslims must look directly for the first visible crescent of the New Moon in order to regulate their religious calendar; they use direct observations to keep track of the beginnings of the months and to work out their year.

For determining the date of Passover, the Hebrews followed the Mosaic rule set down in Exodus (12: 1-8), Numbers (28:16), and Leviticus (23:5): "In the first month, on the fourteenth day of the month, at twilight, there shall be a passover offering to the LORD." The 14th night of the first month equates to the first Full Moon of the year, since the Full Moon always occurs 14 days after the New Moon.

Note that, by this definition, the celebration of Passover was not to them a "moveable feast," but it was fixed as being on the same day of the same month every year: the day of the first Full Moon of the year. But the choice of which month was the first month of the year was proclaimed on the authority of the Sanhedrin, as noted above, who might or might not decide that it was necessary in a given year to add an intercalary month. Thus it was quite possible that they could get it wrong, given the difficulties noted above and the possible inexperience of those making the decision.

Once the New Moon was directly observed by the Sanhedrin, there followed a solemn proclamation, accompanied by the sound of trumpets, sacrifices in the Temple, and the sending forth of messengers to the other cities throughout the country. If bad weather prevented any direct observations, they would proceed using calculations based on the observations made in the previous month. The same procedure was used for the proclamation of the beginning of the year.

After the destruction of Jerusalem in A.D. 70 and the subsequent Diaspora of the Jewish people, it was no longer possible to maintain this system; there was no longer a Temple authority that could announce the beginning of the month or the year. Instead, finally, they decided to adapt the Metonic cycle of 19 years to calculate a calendar, which therefore was no longer based on the direct observation of the Moon. At least in this way the whole Jewish community of the Diaspora could celebrate Passover on the same day.

Above: "The first apparition of the crescent moon was announced to the people by the high priest and proclaimed at the sound of the trumpets...." This rather fanciful illustration is taken from p. 137 of C. Flammarion's Astronomie Populaire.

The Christian Calendar and the Council of Nicea

At first the early Christians followed the Jews in computing the day of Passover, but soon they parted ways, delaying the celebration of Easter to the day immediately after the Sabbath — Sunday — following Passover. This feast had a different significance for the Christians: not the liberation from the slavery of Egypt, but the promise of the Resurrection (which had occurred on a Sunday) and the liberation from sin. But, like the Jews, the various Christian churches were located far from one another and not always in good communication; they each found it necessary to work out for themselves the date for the celebration of Easter. This was a challenge to Christian unity, ultimately confronted by the Council of Nicaea, which met in A.D. 325, an important date in the Christian Calendar.

By that time the calendar of Julius Caesar was in use throughout the Roman Empire. In this calendar you no longer needed to determine when you should insert a month or not, in order to keep the first month in spring; the lunar months of 29 or 30 days were replaced by 12 months that were merely intervals of time set, by convention, to 30 or 31 days. Of course, the lunar months were gone forever in the Roman calendar. The advantage of this system lay in the fact that the Julian year was very well harmonized with the seasons and the motions of the Sun. Given the universal use of this calendar, the council fixed the equinox of the Julian calendar at the time of the council, and then interpret-

ed the text of the Old Testament defining Passover (and thus Easter) in the following mode: one would celebrate Easter on the Sunday that followed the first Full Moon after the 21st of March. If the Full Moon itself fell on a Sunday, Easter would be celebrated on the following Sunday. This was to avoid the confusion of the Christian Easter with the Hebrew Passover. Notice that, by this definition, the calendar no longer depended on an authority who would

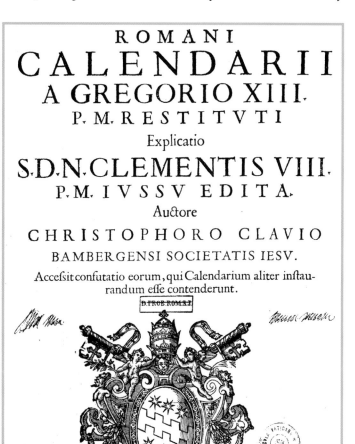

ROMANI
CALENDARII
A GREGORIO XIII.
P. M. RESTITVTI
Explicatio
S.D.N.CLEMENTIS VIII.
P. M. IVSSV EDITA.
Auctore
CHRISTOPHORO CLAVIO
BAMBERGENSI SOCIETATIS IESV.
Accessit confutatio eorum, qui Calendarium aliter instaurandum esse contenderunt.

ROMAE, Apud Aloysium Zannettum. MDCIII.
EX SVPERIORVM PERMISSV.

Above: After Pope Gregory XIII approved the proposed reform of the calendar, the Jesuit priest Christoph Clavius wrote a book explaining the reasoning behind the reform, and how it would work. Shown here is the cover page of a copy from the library of the Vatican Observatory in Castel Gandolfo.

solemnly announce a date for Passover every year, with all its attendant arbitrariness or uncertainty. Now each Church, even those most distant, could calculate for itself when Easter would be celebrated.

Note that the council could have defined Easter in terms of a direct observation of the Full Moon; but instead, they preferred to work out the day of the Full Moon independently of direct observation, because it would not always be possible to do make such observations given the particular meteorological conditions. Instead, they relied on the Metonic cycle described above.

The fathers of the Council of Nicaea were aware of the fact that the position of the date of Easter was merely a mathematical interpretation of the cited texts of the Old Testament. It was left to the Church of Alexandria, a city that had a strong astronomical tradition, the job of preparing a more specific scheme for computing the date of Easter. The system finally adopted for determining the date of the New Moon was worked out some 200 years later by Dionysius Exiguus, a monk who lived in Rome in the first half of the sixth century. In the process, he introduced the concept and numbering of the "Anno Domini" years now in common use.

Dionysius constructed a table that allowed one to read off the date of the New Moon for each month of every Julian year in the 19-year Metonic cycle; this table was repeated every 19 years. The first cycle was said to commence with the year A.D. 1. A *Golden Number* was defined as the value that indicated where the current year sat in the Metonic cycle.

Thus, to compute the date of Easter of any given year, one first determined the Golden Number of the year; then, in Dionysius' table of New Moons, one looked for the first New Moon after March 21. The Paschal Moon, the Full Moon, would occur 13 days later, that is, on the 14th day of the lunar month. The date of Easter would be the Sunday after this Full Moon unless this Full Moon fell on a Sunday; in that case, Easter would be celebrated on the following Sunday. (Knowing which day was Sunday for any given month of any given year could also be computed easily with further tables provided by Dionysius. The sequence of the days of the week continued uninterrupted through all of these calendar reforms.)

Thanks to these New Moon tables, the date of Easter could be computed without ambiguity. The Pascal New Moon would occur between March 8 and April 5, and so the Full Moon could occur from March 21 to April 18. Given the possibility that April 18 itself might be a Sunday, the dates on which Easter could occur thus ranged from March 22 until April 25.

Recall that for the ancient Hebrews, Passover was always fixed on the 14th day of the first month of the year, i.e., the day when the Full Moon actually occurs. For them, the date could be determined without needing to use any complicated, approximate tables; anyone can see the Full Moon, which

was what was particularly important to them for determining holy days. For the Christian Easter, however, this no longer held; indeed, Easter could occur as much as seven days after the Full Moon.

The Gregorian Reform of the Calendar

Surely the fathers of the Nicene Council realized that they were not leaving us a perfect calendar. They understood that the duration of the Julian solar year was slightly too long, and that the difference would be perceptible within one or two generations, as the error would continue to accumulate. Everyone understood that this "defect" introduced an error of about one day in 133 years if one compared the Julian year of 365 (and a quarter) days and the actual value of the year. It wasn't long before voices were raised demanding a reform of the calendar.

By the Middle Ages, calendars were used to indicate the "true" or astronomical day of the equinox, which was the entrance of the Sun onto the first point of Aries, and the "official" equinox, which was always kept on the 21st of March. By the 1500s this difference had already grown to be 10 days. But this probably wasn't the only motive for the reform of the calendar. There was also an error of four days in the determination of the New Moon according to the Metonic cycle, which made the determination of Easter no longer in accord with the spirit of the Council of Nicaea. The situation was getting out of hand.

First of all, one needed to adopt a value of the year that came much closer to its actual length. When Julius Caesar, following the advice of the astronomer Sosigenes of Alexandria, introduced the year of 365.25 days, it was already known that this length was slightly

too long — producing an error of one day in about 133 years. But to make a simple and practical calendar, it was decided to merely intercalate, or insert, one day every four years and leave any further corrections to the distant future. This small difference had accumulated with the passage of time, however, and by the Middle Ages it was evident that the spring equinox no longer coincided with the official equinox of March 21.

Church councils, particularly those held in Constance (1414-1418) and Trent (1545-1563), requested that the Popes work towards finding a correction to the calendar. Their delay in doing so, however, was not due to negligence on the part of the Popes, but because no one had yet presented a reform that was valid and simple, unambiguous, and fully in accord with the Council of Nicaea.

Pietro Pitati, in a treatise published in Verona in 1560, noted that three times 133 years was practically 400 years. Therefore, he realized, a better approximation could be achieved by dropping three days every 400 years: he suggested keeping the regular year of 365 days with a leap year every fourth year, except for years ending with two zeros. But when such years were divisible by 400, it would stay a leap year. (Thus, while 1896 and 1904 were leap years, 1900 was not; but 2000 was.)

But this was only the first step in reforming the calendar. By then the New Moons in Dionysius' table were already off by four days in comparison with actual observed position of the Moon. Pitati studied the lunations with

Opposite: Pope Gregory XIII issued the modern reform of the calendar in 1582.

the astronomical tables of his time, but he could not come up with any truly practical solution. It was left to Aloysius (Luigi) Lilio (1510-1552), a professor of medicine at the University of Perugia, to give the definitive solution.

Recall that Dionysius's scheme, in which one looked up the New Moon for each Golden Number corresponding to the given year, only allowed one to insert extra months as needed to make the lunar and solar calendars

GREGORIVS · XIII · P · M ·

agree. Lilio's idea was to make use of the *epact*, the difference between the lunar and solar year as counted in days, instead of months. This allowed one to adjust the calendar by a day instead of by a whole month. The epact for a given year was defined as the age of the Moon, i.e. the day of the lunar month, on the first day of January of that year. For example, if on this day the Moon was in its 14th day (Full Moon), the epact of that year would be "14." Then one could replace the ancient tables of Dionysius with an equivalent table that replaced the Golden Number with the epact. Then, to compute the date of Easter, one proceeded more or less as before; you used the epact of the year as the Golden Number. And as the Metonic cycle drifted away from the actual occurrence of the New Moon, the epact could be adjusted by the number of days needed to bring the calendar back into agreement with observations. This intercalation was called the *equation* (using an old sense of the word) of the epact.

The "equation" rule finally adopted was to reduce the age of the Moon by one day every 300 years, repeating this over eight such 300-year periods, and then make another adjustment of a day after an interval of 400 years. This 2,800-year cycle could be repeated indefinitely.

Finally, it was decided to make all these intercalations and corrections occur only in years that end with two zeros. These years are called *centenary* years since they begin the century (or end the century, however you prefer to put it). Thus, on each centenary year one might introduce an equation of one day if necessary (shifting back the equation of the Moon every 300 or 400 years) to adjust the Lunar year, while at the same time one was making the appropriate correction (leap year or not) to the solar year. The sum of these two corrections served to modify the tables of the epact, which would take the place of the appropriate Golden Number tables.

Aloysius Lilio's proposal, presented after his death to Pope Gregory XIII by his brother Antonio, was immediately accepted. A Calendar Commission was named to prepare a description of the proposal, called the Compendium, and in 1577 it was sent to all the civil authorities of Europe, including universities and academies. After going through all the responses, the Calendar Commission then prepared the Papal Bull, *Inter Gravissimas*, which in 1582 decreed the adoption of the new calendar.

Arguably, the introduction of the new calendar was in fact more of a correction than a reform. One merely used a slightly more accurate value for the length of the year and likewise a slightly adjusted method for calculating the phases of the Moon. The Julian rule for leap years was only somewhat modified. With Lilio's mechanism of epacts, the Metonic cycle was preserved as a valid way to calculate Easter in accord with tradition. Finally, for one time only, 10 days were eliminated to move the equinox back to March 21, bringing it back into accord with the official date in use since the Council of Nicaea. This was the great merit of the Gregorian reform: with a minimum of adjustment and with clear rules, it preserved the intent of the council. And because it relied on a council that had sat before the schism between East and West, it was hoped that this reform could avoid further conflict with the Orthodox Church.

An important member of this commission for the reform of the calendar was Fr. Christoph Clavius, S.J., a professor of mathematics at the Roman College, known for his publications in geometry, arithmetic, and astronomy. It is difficult to know for sure his full role in the commission, but it is enough to note that he was the one instructed by the Pope to describe and defend the new calendar. His work, *Explanatio Romani Calendarii* (Rome, 1603), is fundamental and indispensible for anyone studying the reform.

As he pointed out, there were any number of other possible solutions for the reform of the calendar. For example, one could have left the true equinox at March 11, where it was at that point; or completely simplify the whole question of finding Easter by fixing it to a non-movable date; or perhaps use astronomical tables to determine the first Full Moon of spring. But these would have been a significant break with the tradition that had followed the text of the Old Testament to determine the date of Passover. It was preferred instead to respect as much as possible the spirit of what had been determined at the Council of Nicaea and the longstanding tradition of the Church. Since the true equinox had originally been set at March 21, that would not be touched in the rule for determining Easter. In addition, one wanted a simple system in which there would be no need for a special understanding of astronomy, in a format that could be easily transported and adapted to distant regions by explorers and missionaries, who would be able to generate the calendar accurately and without ambiguity.

In Chapter IV of his book, Clavius gave additional reasons for adopting this version of the calendar. He maintained that if one were to use the true astronomical value for the New Moon instead of the approximate value according to the Metonic cycle of 19 years, this would likely just provoke arguments and so put Christian unity (already strained by the Protestant Reformation) further at risk. Using astronomical tables would not help, because of the discrepancies among the diverse tables available at that time. Instead, it seemed more convenient to adopt a calendar that anyone could carry with them and use, rather than rely on experts who were not always in agreement. That was certainly the case in the 16th century.

The new calendar was accepted almost immediately in Catholic countries. There was, however, a great reluctance to adopt it in the Protestant lands, for obvious political and religious reasons. Some Protestant communities in northern Germany chose instead to calculate the dates of the Full Moon in accord with published ephemerides; this recourse to astronomical data, as noted by Clavius, gave them an excuse for evading the Gregorian reform. Only at the beginning of the 18th century was the reform accepted by all of Europe except the Orthodox Christians. (A complete table of the dates of adoption in each country can be found in the *Explanatory Supplement to the Astronomical Ephemeris*, prepared by the Royal Observatory in London; or see E. G. Richards, *The Mapping of Time*, Oxford, 1999.)

people thought that the end of the world would occur long before this day arrived. In any case, they reasoned, it should not be all that hard to make the tiny adjustment of one day necessary after so many years. As it happens, however, comparing the Gregorian year with the modern tropical year, one finds that the difference of one day actually will come much sooner, after 3,000 years.

Still, there is really no point in looking for a way to improve the Gregorian tropical year. Every so often, someone points out that the value of the Gregorian year is in error; but generally what they forget is that the reform commission was well aware that they had not adopted the best and most precise value of the tropical year then available. It was, instead, a value conditioned by the decision to make the interpolation only during centenary years.

Modern Attempts to Reform the Gregorian Calendar

We move now to the attempts of various societies and nations to come up with a perpetual calendar. The rock on which every such proposal has foundered is that there is not an exact number of weeks in a year. There is always one day extra, two in leap years, that does not belong to any week; and many religious groups absolutely refuse to accept any interruption in the weekly cycle (thus interrupting the celebration of the Lord's Day).

Other proposals have been made to adjust the value of the length of the Gregorian year, which is slightly longer than the true value; but they forget that the day is also variable, as we will show below, and it is not advisable to modify the year to obtain a better match to the fractional number of days

Above: Fr. Christoph Clavius, S.J., a professor of mathematics at the Roman College, known for his publications in geometry, arithmetic, and astronomy; he was instructed by the Pope to describe and defend the new calendar.

According to the reform commission, the new calendar was in essence perpetual, in the sense that if one compared the value of the tropical Gregorian year with the value found in the Alphonsine table of 365.24255 days, you would find an error of only one day in about 20,000 years. Many

in a year, when you don't know exactly what the length of the day itself will be in future times.

At a gathering of the World Council of Churches held in Aleppo, Syria, in 1997, it was proposed that the Gregorian calendar's rule for the determination of Easter should be eliminated in favor of a more precise calculation, more in keeping with improved astronomical ephemerides. That had already been tried for a while in various Protestant countries, until they eventually decided to adopt the Gregorian calendar; and certainly, the editors of printed calendars today already include the dates of the phases of the Moon based on astronomical almanacs. But while this might seem to be a reasonable idea, in practice it is not so easy or obvious to implement when it comes to determining the date of Easter.

First of all, if you are going to use astronomical tables, you have to choose which longitude you will use for your computations (for example, to determine whether it is Saturday or Sunday when a given Full Moon occurs). Let's say, by mutual agreement, you choose to make your calculations by assuming you are in Jerusalem. Now, while the "spring equinox" for the Gregorian calendar is always fixed at March 21 independent of the actual position of the Earth, it can happen that spring actually starts when it is March 20 or even, some years, March 19 at Jerusalem's longitude. In preparation for the Aleppo conference, the date of Easter was calculated using both the astronomical and Gregorian calendar methods. From the year 2001 through 2100 the astronomical and Gregorian methods disagreed nine times. It is not surprising that there was such a difference, because the use of the Jerusalem longitude influences all the dates of the equinox used in determining which Full Moon is the first Full Moon of spring.

436 CALEND. GREG.
TABVLA FESTORVM

Anni Domini	Aur. Num.	Epactæ	Literæ Dominicales Calend. Gregor.	Plenilun. media Caléd. Gregoriani (D. H.)	Lunæ xiiij. Calendarij Gregoriani	Septuagesima	Dies Cinerum	Pascha Calend. noui	Ascésio Domini
1984	9	xxvij	A g	15.10.A	16. A	19. Feb.	7.Mar.	22. A	31.Maij
1985	10	viij	f	4.19.A	5. A	3. Feb.	20. Feb.	7. A	16.Maij
1986	11	xix	e	25. 4.M	25. M	26. Ian.	12. Feb.	30. M	8.Maij
1987	12	✳	d	13. 2.A	13. A	15.Feb.	4.Mar.	19. A	28.Maij
1988	13	xj	c b	1.10.A	2. A	31. Ian.	17. Feb.	3. A	12.Maij
1989	14	xxij	A	21.19.M	22. M	22. Ian.	8. Feb.	26. M	4.Maij
1990	15	iij	g	9.17.A	10. A	11.Feb.	28 Feb.	15. A	24.Maij
1991	16	xiiij	f	30. 2.M	30. M	27. Ian.	13.Feb.	31.	9.Maij
1992	17	25	e d	16.23.A	17. A	16. Feb.	4.Mar.	19. A	28.Maij
1993	18	vj	c	6. 8.A	7. A	7. Feb.	24. Feb.	11. A	20.Maij
1994	19	xvij	b	26.17.M	27. M	30. Ian.	16. Feb.	3. A	12.Maij
1995	1	xxix	A	14.14. A	14. A	12.Feb.	1.Mar.	16. A	25.Maij
1996	2	x	g f	2.23.A	3. A	4.Feb.	21.Feb.	7. A	16.Maij
1997	3	xxj	e	23. 8.M	23. M	26. Ian.	12.Feb.	30. M	8.Maij
1998	4	ij	d	11. 5.A	11. A	8. Feb.	25.Feb.	12. A	21.Maij
1999	5	xiij	c	31.14.M	31. M	31. Ian	17.Feb.	4. A	13.Maij
2000	6	xxiiij	b A	18.12.A	18. A	20. Feb.	8.Mar.	23. A	1. Iun.
2001	7	v	g	7. 21.A	8. A	11 Feb.	28.Feb.	15. A	24.Maij
2002	8	xvj	f	28. 5.M	28. M	27. Ian.	13.Feb	31. M	9.Maij
2003	9	xxvij	e	16. 3.A	16. A	17.Feb.	5.Mar.	20. A	29.Maij
2004	10	viij	d c	4.21.A	5. A	8. Feb.	25.Feb.	11. A	20.Maij
2005	11	xix	b	24.21.M	25. M	23. Ian.	9. Feb.	27. M	5.Maij
2006	12	✳	A	12.18.A	13. A	12.Feb.	1.Mar.	16. A	25.Maij
2007	13	xj	g	2. 3.A	2. A	4. Feb.	21.Feb.	8. A	17.Maij
2008	14	xxij	f e	21.12.M	22. M	20. Ian.	6. Feb.	23. M	1.Maij
2009	15	iij	d	9. 9.A	10. A	8. Feb.	25.Feb.	12. A	21.Maij
2010	16	xiiij	c	29.18.M	30. M	31. Ian.	17.Feb.	4. A	13.Maij
2011	17	25	b	17.16.A	17. A	20. Feb.	9.Mar.	24. A	2.Iun.
2012	18	vj	A g	6. 1.A	7. A	5. Feb.	22.Feb.	8. A	17.Maij
2013	19	xvij	f	26. 9.M	27. M	27. Ian.	13.Feb.	31. M	9.Maij
2014	1	xxix	e	14. 7.A	14. A	16.Feb.	5. Feb.	20. A	29.Maij
2015	2	x	d	3.16.A	3. A	1.Feb.	18.Feb.	5. A	14.Maij
2016	3	xxj	c b	23. 0.M	23. M	24. Ian.	10. Feb.	27. M	5.Maij
2017	4	ij	A	10.22.A	11. A	12.Feb.	1.Mar.	16. A	25.Maij
2018	5	xiij	g	31. 7.M	31. M	28. Ian.	14. Feb.	1. A	10.Maij
2019	6	xxiiij	f	19. 4.A	18. A	17.Feb.	6.Mar.	21.	30.Maij

In addition, while the Gregorian calendar uses a simple, easy-to-use arithmetic based on whole numbers, with the proposed change you would have to use fractional ephemeris values that are constantly varying. Normally, this would not make any difference between the Easter determined by the Gregorian and the "astronomical" calendar; but if it should happen that the Full Moon occurs a few minutes before or after midnight (in Jerusalem), this could lead to problems. Fortunately, this wouldn't occur very often, but it cannot be ruled out: there would be uncertain cases if you wanted to make any long-term calculations for the date of Easter.

The reason for this situation is that lunations are calculated with what is called "Terrestrial Dynamic Time" (which is found, in practice, with "atomic clocks": very precise clocks based on various properties of atoms). But the lunations that are needed for the liturgical computation of Easter are calculated in "Universal Time," which is obtained simply by counting the number of days — i.e., the number of rotations of the Earth about its axis. Well, the difference ΔT between these two times is slightly variable in an unpredictable way, which changes over time, due to the fact that the rotation of the Earth about its axis is not uniform. Even if we could calculate in Terrestrial Dynamic Time the precise position of the Sun and the Moon for some far future time, that would not be the position that would be used by a calendar, which is measured in Universal Time. Therefore it would

not be possible to prepare a table of Easter dates for the distant future, but only for an interval of years for which the unknown ΔT is not critical; in other words, so long as we don't have the situation — very rare, to be sure — when the New Moon occurs within the error of the ΔT of the ephemeris. Naturally this influences only the calculation of Easter for the far distant future, but it would be necessary to create an authority to decide what to do in such special circumstances.

The Gregorian rule for the calculation of Easter is in practice very simple and proceeds automatically without needing any authority that would have to make decisions in uncertain cases, since there are no such cases. After 3,000 years, there will be the need to eliminate one day in the year because, for reasons of simplicity as we have described, the Gregorian year is still a little too long. The beginning of a millennium can provide a perfect opportunity to correct the year and at the same time make an adjustment of a whole number of days in the equation of the New Moon, in a way analogous to the fundamental rule of the Gregorian reform of the calendar.

The introduction of a more "scientific" determination of the spring Full Moon does not affect a fundamental problem of the date of Easter: when should it occur when the choice lies at the extremes of the season, and one must choose either March 22 or, say,

April 25? In other words, should Easter be celebrated at the beginning of spring or in roughly the middle of the season? This oscillation of the date of Easter carries with it the dates of all the other movable feasts, such as the beginning of Lent, the feast of Pentecost, and the Sundays of Ordinary Time during the year. Of course, fixing Easter independently of the Full Moon — for example, at the last Sunday of March or the first Sunday of April — would reduce this oscillation to less than one week. But obviously this would run contrary to the spirit of the Council of Nicaea.

On the whole, it would not seem to be a good idea to introduce changes into the calendar unless they have been well thought-out and are likely to be enduring. If nothing else, such changes would lead to great confusion in historical chronology. Above all, whatever the modifications of the rules of the calendar, the date of Easter in particular has a fundamental importance for Christians, who would want to act in accord with all of the other Christians to avoid further divisions and confusion. One can well ask if the increase in precision that would come with the introduction of using lunar positions calculated with a modern ephemeris would be justified, if it otherwise did not follow either the Old Testament or the traditional practice of Christians. In any case, one should not forget that the principle intent that guided the Council of Nicaea was not so much astronomical precision as the unity of all Christians in the celebration of Easter. ●

Left: The power of the Gregorian Reform is that one can calculate dates indefinitely into the future. This table, taken from Clavius' book on the Gregorian Reform, lists the dates of movable feasts like Easter (Pascha Calend. noui — Easter, New calendar — second column from the right) calculated more than four hundred years into the future.

FR. JUAN CASANOVAS, S.J. (Spain), is a solar astronomer and historian of astronomy. This chapter was adapted and translated by Guy Consolmagno from an article, "La Determinazione della Pasqua," published in the journal Rivista Liturgica in 2001.

GALILEO AND HIS TIMES: SOME EPISODES

• Fr. GEORGE V. COYNE, S.J. •

Galileo's Times

Galileo lived in exciting times, both for himself, for the Church, and for humanity in general. His view of the cosmos became a celebrated controversy, to a great extent, because of the times in which he lived. And his view of the cosmos was, to some extent, an inheritance from times gone by, as far back as the ancient Greek Aristarchus. So the past and the present for Galileo are very much intertwined.

What were some of the ideas that were circulating in Galileo's time? Until Galileo's epoch, discussions as to whether the Earth or the Sun is at the center of what we now know as our Solar System was a topic of placid debate and not of heated controversy, essentially because any claims that one or the other represented the real universe were not at issue. Both proposals were seen as only hypothetical systems, mathematical expedients to manage what was observed in the heavens.

This sense of the hypothetical had a long history to it, going back to the time of the Pythagoreans five centuries before Christ. There was no thrust to understand how the heavens really worked, as long as we had a mathematical technique to predict where objects that moved in the sky would be found.

To continue this way of thinking, of course, would be a betrayal of what scientists are really trying to do; and Galileo was going to redeem that betrayal. Galileo would completely undermine these ideas. He would provide persuasive, although not conclusive, evidence for a Sun-centered system. And in so doing, he would topple the classical Greek philosophy of nature, which had dominated thinking about the universe for millennia.

Galileo, Teacher and Scientist

What was Galileo's personal life like? After attempts to obtain a teaching position at Bologna, Padua, and Florence, in July 1589 Galileo was called to a teaching position at Pisa. He taught the elements of mathematics and astronomy.

What were the sources for Galileo's teaching? Historians working during the past decades have concluded that Galileo relied to a great extent upon lecture notes of Jesuits at the Roman College; that is where Galileo came in contact with the Aristotelian natural philosophy: what today we call physics.

For Aristotle there were only four elements: earth, air, fire, and water. Since earth was the heaviest, it was taught, it had to sink down to the center. Furthermore, all heavenly bodies were perfect, so they had to be unblemished spheres that moved in perfect cir-

Above: While visiting Italy in the mid-1820s, J.M.W. Turner painted this watercolor of the Moon shining on Galileo's villa outside Florence.

GALILÆUS GALILÆI LYNCEUS

PHILOSOPHUS et MATHEMATICUS

Ser:ᵐⁱ Hetruriæ magni Ducis.

J. Mulder Fecit.

Above: *A portrait of Galileo from a 1699 edition of his controversial book,* On the Two World Systems, *which defended the Copernican theory. (From a copy in the Vatican Observatory library in Castel Gandolfo)*

centered system in astronomy, at least for didactic purposes. But both they and Galileo shared in the growing tensions between an Aristotelian natural philosophy and the new scientific discoveries, especially those of Galileo soon to appear. For the Jesuits, this would create an even more significant tension in the realm of theological and doctrinal issues, since the latter relied heavily upon a "Christianized" Aristotelian philosophy.

There is a personal relationship involved in this connection of Galileo, the teacher from Pisa, with the Roman College. In 1587 Galileo took a trip to Rome for his first meeting with the famous Jesuit mathematician, Christoph Clavius, to seek a recommendation from him for a teaching chair in mathematics at Bologna. (He got the recommendation, but not the job.) This provided Galileo with an opportunity to see firsthand the teaching of the philosophy professors at the Roman College. It was then through his regular correspondence with Clavius that Galileo would have obtained the various teaching notes from the Roman College, which he adapted to his own teaching at Pisa.

When Galileo first visited him, Clavius was already at the height of his fame. Clavius was then 50 years old; he had single-handedly founded the world-renowned school of mathematics at the Roman College, and had published various treatises in mathematics and astronomy, which had a wide circulation. He had played an important role in the reform of the calendar under Pope Gregory XIII. Galileo, who was 27 years his junior and had just begun his scientific career, impressed Clavius with his talent both in theory and in practical matters. A personal relationship based upon a deep esteem for each other was born at that time and it lasted, despite

cular motions. This was the conceptual approach that Galileo inherited and which he would question as his thinking matured. He would eventually completely undermine these ideas.

The Jesuits at the Roman College undoubtedly followed Aristotle in philosophy, and they taught an Earth-

some travail, until Clavius' death in 1612.

After Pisa, Galileo began teaching in Padua, and as he himself said, he spent the happiest 18 years of his life there. Padua was part of the Venetian Republic, which at that time found itself on various issues in opposition to Rome. The Jesuits were the defenders of Papal authority, and several of Galileo's friends, defenders of the independence of the Venetian Republic, found themselves in opposition to the Jesuits. This, undoubtedly, had some influence on Galileo's attitude towards the Jesuits, but it is also clear that Galileo maintained a cordial and productive relationship with Clavius and his disciples at the Roman College.

Galileo's Discoveries
with the Telescope

It was during his time in Padua that Galileo carried out his epoch-making telescopic discoveries. In November of 1609, he finally succeeded in making a telescope that magnified 20 times. With it, he observed that the Moon has blemishes, mountains and craters; Venus has phases; there are four satellites going around Jupiter; there are myriads of stars in the belts of light traversing the sky known as the Milky Way.

Before we turn our gaze upon Galileo with his telescope pointed to the heavens, I would like to attempt to recover his state of mind at that moment. He was nearing the end of a relatively long, tranquil period of teaching and research, during which he had come to question at its roots the orthodox view of the known physical universe. But he had as yet no solid physical basis upon which to construct a replacement view. He sensed a unity in what he experienced in the laboratory and what he saw in the heavens. But his view of the heavens was limited, although there was some expectation that, since with his telescope he had seen from Venice ships at sea at least 10 times the distance at which they could be seen by the naked eye, he might go a bit beyond that limit when looking into the sky.

He was uncertain about many things in the heavens. For example, he had seen an object suddenly appear as bright as Jupiter and then slowly disappear (what we now know as an exploding star, a supernova). He had been able to conclude that it must be in the realm of the fixed stars. But he could venture nothing about its nature.

Did he expect that the telescope would let him find out for certain whether the Earth was going about the Sun? Probably not; his expectations were not that specific. He simply knew that the small instrument he had worked hard to perfect, since he had already convinced his patrons of its value for military purposes, was surely of some value for scientific purposes. Although it may seem obvious to us, that in itself was a major discovery: that one could learn more about nature by building artificial instruments like a telescope. For the first time, our knowledge of the universe would be shaped by more than what anyone could simply see with the unaided eye.

In brief, I propose to you a Galileo who was extremely curious, anxious to resolve some fundamental doubts, and clever enough to know that the instrument he held in his hands might contribute to settling those states of mind.

Obviously, not everything happened in the first hours or even the first nights of observing. The vault of the heavens is vast and varied. It is difficult to reconstruct in any detail the progress of Galileo's observations; but from November 1609 through January 1610 there is every indication that he was absorbed in his telescopic observations. From his correspondence we learn that he had spent "the greater part of the winter nights under a peaceful open sky rather than in the warmth of his bedroom." They were obviously months of intense activity, not just at the telescope but also in his attempt to absorb and understand the significance of what he saw. His usual copious correspondence becomes significantly reduced during these months, but we do learn from it that he continued in his attempts to improve his telescope.

At times his emotional state breaks through in his correspondence. He makes a climactic statement in this regard in a letter of January 20, 1610, some weeks after his observations of the moons of Jupiter, when he states: "I am infinitely grateful to God who has deigned to choose me alone to be the first to observe such marvelous things which have lain hidden for all ages past." For Galileo, these must have been the most exhilarating moments of his entire life. The observations were carefully recorded in *The Starry Message*

mosaic? He actually said little of any scientific significance about this in *The Starry Message;* and rightly so, since his observations had gone far beyond his capacity to understand. But he could still marvel. By contrast he showed a very acute insight when it came to sensing the significance of his observations of the Moon, of the phases of Venus, and, most of all, of the moons of Jupiter. The preconceptions of the Aristotelians were crumbling before his eyes.

He had remained silent long enough. Over a three-month period he had contemplated the heavens. It was time to organize his thoughts and tell what he had seen, and what he thought it meant. It was time to publish!

It happened quickly. The date of publication of *The Starry Message* can be put at March 1, 1610, less than two months after his discovery of Jupiter's brightest moons and not more than five months after he had first pointed his telescope to the heavens. With this publication, both science and the scientific view of the universe were forever changed. For the first time in over 2,000 years, new significant observational data had been made available to anyone who cared to think not in abstract preconceptions but in obedience to what the universe had to say about itself.

Did Galileo's telescopic discoveries prove that the Earth went about the Sun? Did Galileo himself think that they had so proven? There is no simple answer to these questions, since there is no simple definition of what one might mean by proof. Let us limit ourselves to asking whether, with all the information available to a

but, by necessity, stripped for the most part of their emotional content. What must have been, for instance, Galileo's state of mind when he viewed for the first time the Milky Way in its entire splendor: innumerable stars resolved for the first time, splotches of light and darkness intertwined in an intriguing

Above: Galileo's handwritten manuscript describing his discovery of the moons of Jupiter using his new telescope. This text was prepared in 1610 for his first astronomy book, The Starry Message. *(From a facsimile reproduced in a complete set of Galileo's writings, published in Florence in 1892.)*

most stupid and silly populace, the witness of the stars themselves would not be enough, even if they came down to the Earth to tell their own story." While he could not bring the stars to Earth, he had, with his telescope, taken the Earth towards the stars. He would spend the rest of his life drawing out the significance of those discoveries.

contemporary of Galileo's, it was more reasonable to consider the Earth as the center of the known universe or that there was some other center. The observation of at least one other center of motion (moons moving around Jupiter instead of the Earth); the clear evidence (in the crater-pocked face of the Moon and spots on the Sun) that at least some heavenly bodies were "corrupt"; and most of all, the immensity and density of the number of stars which populated the Milky Way — these left little doubt that the Earth could no longer be reasonably considered the center of it all.

Galileo's own convictions are clear, for instance, from his own statement in the Dialogue: "... if we consider only the immense mass of the sphere of the stars in comparison to the smallness of the Earth's globe, which could be contained in the former many millions of times, and if furthermore we think upon the immense velocity required for that sphere to go around in the course of a night and a day, I cannot convince myself that anyone could be found who would consider it more reasonable and believable that the celestial sphere would be the one that is turning and that the globe would be at rest."

But Galileo was also wise enough to know that not everyone could be easily convinced. In a letter to a friend he wrote: "... to convince the obstinate and those who care about nothing more than the vain applause of the

Galileo and the Jesuits at the Roman College

Riding on the crest of his telescopic observations, Galileo planned a trip to Rome. Cardinal Bellarmine had heard of Galileo's observations and wished to know if they were true, and what implications they held. Bellarmine turned to his fellow religious, the Jesuits at the Roman College, and asked them to test Galileo's observations.

The day after his arrival on March 29, 1611, Galileo paid a long and cordial visit to the Jesuit astronomers and mathematicians. He was honored by an academic assembly at the Roman College with the participation of numerous cardinals and other personages of Roman Society. The official oration, entitled "Starry Message of the Roman College," clearly alluded to Galileo's book of celestial discoveries; it lauded Galileo for his observations and announced that they had been confirmed unanimously by the Jesuit astronomers and mathematicians at the College.

On the other hand, they were cautious about discussing the observations in terms of a Sun- or Earth-centered system. It is clear from statements of Clavius and those of his Jesuit colleagues at the Roman College at that time that they were hesitatingly approaching Copernicanism, a Sun-centered system. The hesitation was shared by the Jesuit philosophers and theologians of the Roman College, who were not pleased with the all-too-posi-

tive appreciation of Galileo's discoveries and especially the anti-Aristotelian implications of those discoveries. It was reported that the statements of the Jesuit astronomers on the observations of Galileo were accompanied by murmurings on the part of their philosopher colleagues.

The hesitation on the part of the philosophers was due to their need to maintain "uniformity of doctrine." That persistent requirement of fidelity to Aristotelianism had nothing to do directly with a Sun-centered system. It was motivated by the conviction that Aristotle furnished a solid basis for philosophy and, upon adaption, for the so-called "preambles of the faith." But now the natural philosophy of Aristotle was crumbling. Would his whole system of philosophy crumble also — and take with it the whole system of medieval theology that it supported?

The structure of the Aristotelian system was a unified whole. If the natural philosophy of Aristotle crumbled, would the structure itself give way? How then to maintain "uniformity of doctrine"? There was not, of course, an open, public schism among the philosophers, mathematicians, and astronomers of the Roman College. Loyalty to a tradition, reinforced by religious superiors, remained the dominant factor. But the Jesuit astronomers were steadily embracing Copernicanism and thereby threatening Aristotelianism.

The case of one Jesuit cardinal, Robert Bellarmine, is an interesting one. In his early years of teaching at Louvain, he had shown a very independent view of Aristotle. He did not hold, for instance, that the heavens ent to Catholic doctrine was up for grabs. For Bellarmine, the issue was that a Sun-centered universe, that of Copernicus and Galileo, appeared to be untenable theologically because it appeared to contradict Scripture.

Above: The frontispiece from a 1699 edition of Galileo's controversial book, On the Two World Systems. *(From a copy in the Vatican Observatory library in Castel Gandolfo.)*

were immutable and incorruptible. As he matured as a Jesuit, it became clear that he was neither a devotee nor an opponent of Aristotelian natural philosophy. With respect to Aristotle, he was an eclectic: whatever supported Catholic doctrine in that natural philosophy was fine; what was indiffer-

Galileo's Troubles with the Church

And so why did Galileo get into trouble? It is difficult to appreciate the conflict that arose from Galileo's research and writings unless we also understand the politics and religion of his time. It may be helpful to gather together some dates: Copernicus (1473-1543, *De Revolutionibus* in 1543); Martin Luther (1483-1546); Council of Trent (1545-1563); Galileo (1564-1642); Thirty Years War (1618-1648); Isaac Newton (1642-1727). Religious and political conflicts were very ripe and intertwined, as Galileo and his colleagues in the birth of modern science came on and off stage.

In its opposition to Martin Luther and the Reformers, the Council of Trent had declared that there was to be no private interpretation of Scripture: Scripture was the Book of the Church, and only the Church could interpret it authentically. Galileo offered his interpretation of Scripture, whereby he essentially anticipated what the Catholic Church was to propose almost three centuries later; but he did so privately. He said (quoting, by the way, a cardinal) that the Scriptures were written to teach us how to go to heaven and not how the heavens go. For him, there was no scientific teaching in Scripture.

But in 1616 the theologians of the Holy Office declared that the proposition that the Sun is the center of the world and immovable was "absurd in philosophy, because it contradicted Aristotle, and heretical, because it contradicted Scripture and the Church Fathers. Furthermore, the proposition that the Earth is not the center but moves is also absurd in philosophy and erroneous as to Catholic doctrine."

Galileo was not on trial, but he was clearly a target of the Holy Office. This becomes clear when, on the day after this declaration of the theologians of the Holy Office, it was reported at the weekly meeting of the cardinals of the Holy Office that Pope Pius V had requested Cardinal Bellarmine to have a private meeting with Galileo, in which he was to warn him to abandon his views on Copernicanism. Bellarmine did this, and Galileo acquiesced. Within days of these events, the Congregation of the Index published a decree prohibiting or "correcting" a number of writings favoring a Sun-centered system, including those of Copernicus, but Galileo was not explicitly mentioned. These events conclude what is often referred to as the "First Trial."

But with the death of Pope Pius V and the election of Galileo's friend and fellow philosopher Maffeo Barbarini as Pope Urban VIII, it seemed that Galileo would again be free to pursue his ideas. He published several works with Church approval in the 1620s, most notably *The Assayer* (1622), which helped define our modern idea of science. The official church censor, Fr. Nicolo Riccardi, wrote: "I believe our age is to be glorified by future ages … thanks to the deep and sound reflections of this author in whose time I count myself fortunate to be born…."

With all this support, why did Galileo's *Dialogue on the Two Chief World Systems* (1632) eventually lead to his trial and conviction?

The episodes of the Thirty Years War clearly had an impact on the ever-changing relationship between Galileo and Pope Urban VIII. Pope Urban had been a close and longtime friend of Galileo through the earlier years of his papacy. But in 1630, at precisely the time when Galileo was seeking an imprimatur for his famous Dialogue, which eventually led to his condemnation, the conflict between the warring factions in the Thirty Years War began to weigh heavily on Urban VIII. Cardinal Richelieu of France was growing ever stronger in his opposition to Rome. And then in 1632, Cardinal Gaspare Borgia openly and violently attacked the Pope in a consistory of cardinals, with accusations of taking the part of heretics because he had favored an agreement among the King of France, the Duke of Bavaria, and the Protestant King of Sweden in opposition to the Hapsburg agreement between Spain and the German Empire. In brief, the Pope was being accused of betraying the Catholic cause in Europe. These were not the best times for Galileo to seek an approving attitude towards his *Dialogue* from the Pope.

The Trial of 1633, following upon Galileo's publication of his Dialogue in 1632, brought to a culmination the Church's opposition to him. He was declared to have disobeyed the orders given to him in 1616, he was made to abjure his opinions on Copernicanism, and he was sentenced to imprisonment (eventually, house arrest in his home outside Florence).

In the Galileo case, the historical facts are that further research into the Copernican system was forbidden by the decrees of 1616 and then condemned in 1633 by official organs of the Church with the approbation of the reigning Pontiffs. Galileo was a renowned world scientist. The publication of his *Starry Message* established his role as a pioneer of modern science. He had provoked anew the controversy about the local universe. Observational evidence was increasingly overturning Aristotelian natural philosophy, which was the foundation of an Earth-centered system. Even if the Sun-centered system in the end proved to be wrong, the scientific evidence had to be pursued. A renowned scientist, such as Galileo, in those circumstances should have been allowed to continue his research. He was forbidden to do so by official declarations of the Church. That was a tragedy for Galileo, for science, and for the Church.

The Modern Church and the Galileo Affair

What has happened since then to address that tragedy?

With the condemnation of the Church, the Copernican system could not be used as a basis of natural philosophy. But it could still be considered as a useful "hypothesis" in the Pythagorean sense, as a tool for calculating the positions of the planets. Thus, in Jesuit colleges throughout Europe, the study of astronomy — with the idea of a Sun-centered system — quietly moved out of the philosophy classroom and into the mathematics classroom.

The Jesuit theologian and cardinal, Robert Bellarmine, had played a key role in the events of 1616. Bellarmine was not a dyed-in-the-wool Aristotelian, as noted above, but he was profoundly convinced, contrary to the statement of Cardinal Baronio, replayed by Galileo, that "Scripture teaches us how to go to heaven and not how the heavens go," in some instances the Scriptures do teach a natural philosophy. While the personality and high Church office of Bellarmine might tend to dominate any judgment of the role of the Jesuits, he was not necessarily representative of a Jesuit position, if there were such. Probably most representative was that of the Jesuit astronomers of the Roman College, although simplifications are required even here to be able to speak of a Jesuit position.

The Jesuits astronomers were not ivory tower "pure scientists." They lived and breathed a climate of diversity and intellectual intensity with their philosopher and theologian colleagues. They were devoted with the same fidelity to tradition and Church teaching, but they were also participants in the birth of modern science.

Even the preliminary discoveries of that science were challenging the existing basis of Catholic doctrine and the very meaning of Scripture. There was no philosophy of nature to replace that of Aristotle, which was crumbling under the onslaught of astronomical observations. The position of the Jesuit astronomers in general was one of expectation, and certainly not one of timidity or fear: Jesuits taught the mathematics of the Copernican system from Germany to China, and named the most prominent crater on the Moon after Copernicus.

By the 18th century, a hundred years after the Galileo trials, the Copernican system was well known and well accepted in Catholic circles. Isaac Newton's *Principia* (1687) had finally provided an alternative to Aristotle's cosmology that was compatible with the astronomical observations. But the Church's official condemnation of 1633 was not withdrawn until 1757. Galileo's own writings remained on the Index of Prohibited Books until

1835, even though the ideas in them had long since been accepted by the Church. And during all that time, the personal injustice to Galileo himself had never been addressed.

In a discourse of November 10, 1979, in the first year of his papacy, John Paul II spoke of the fact that "Galileo had much to suffer … at the hands of individuals and institutions within the Church." In 1982, the Pope set up the Galileo Commission so that, in his own words: "theologians, scholars, and historians, animated by a spirit of sincere collaboration, will study the Galileo case more deeply…." But in the Pope's discourse of October 31, 1992, which closed the work of the Galileo Commission, the whole Galileo affair is summed up as a "tragic mutual incomprehension" from which a "myth" has continued whereby the Galileo controversy has become a symbol of what some think to be an inevitable contrast between science and faith. Both Galileo and the Church of his time were uncomprehending.

The first discourse seemed to imply that Galileo need not have suffered and that the official Church held some responsibility for his sufferings. In the final discourse, the implication is that Galileo's suffering was inescapable ("tragic" in the sense of the classical Greek tragedies) because of the "mutual incomprehension," inevitable if we consider those times … there is no one responsible for Galileo's suffer-

ings; they had to be; they were "tragic"; they were driven by the uncontrollable circumstances of that historical period.

And so does the myth of Galileo continue.

Galileo's telescopic discoveries surely brought us completely new and unexpected information, and it dramatically overturned the existing view of the universe. It looked to the future. Were there other "centers of motion," such as that seen with Jupiter and its moons? Did other planets like Venus show phases and changes in their apparent sizes? And what to make of those myriads of stars concentrated in that belt which crosses the sky and is intertwined with bright and dark clouds? All of these were questions for the future.

Although neither Galileo nor any of his contemporaries had a well-developed comprehensive theory of the universe, Galileo clearly intuited that what he saw through his telescope was of profound significance. His discoveries were not limited to looking; they involved thinking. Henceforth no one could reasonably think about the universe in the tradition of Aristotle, which had dominated thinking for over two millennia. A new way of approaching the universe was required.

The adventure of scientific discovery was only beginning. Eventually, all else would accommodate itself to what the universe has to say to us. Modern science was being born, and the birth pangs were already being felt. We know all too well how much Galileo suffered in that birth process. ●

FR. GEORGE COYNE, S.J. (USA), is director emeritus of the Vatican Observatory. His research has centered on the study of circumstellar dust. He was a member of the Galileo Commission of the Pontifical Academy of Sciences during the papacy of Pope John Paul II. This chapter was written specially for this book.

A History of the Vatican Observatory

• Fr. Sabino MAFFEO, S.J. •

Above: *A telescope dome of the observatory built on the Tower of the Winds atop the Vatican Library in the late 19th century.*

The Origins of the Specola

The Specola Vaticana (Vatican Observatory) can trace its roots to two different strands of astronomical institutions supported by the Church. Astronomers at the Vatican participated in the Gregorian reform of the calendar in 1582, and the Vatican continued to support astronomy as a part of the Holy See on and off for the next 300 years. Meanwhile, other astronomical research programs were begun by the Jesuits at the Roman College at the time of Galileo.

However, by the 1860s most of the territory of the Holy See had been incorporated into the new Italian state, and in 1870 Rome itself became part of Italy, leaving the Holy See as merely a small enclave around St. Peter's. The expropriation of the Roman College Observatory by the new Italian State in 1879 deprived the Holy See of the last place to carry out astronomical research. This was all the more to be decried because this institute, the home of the Jesuit astronomer Fr. Angelo Secchi (cf. pp. 89-90), had become a world-renowned observatory.

But the Church soon found a new motivation for the support of science.

In 1888, for the jubilee of the priesthood of Pope Leo XIII, a special exhibit was set up of instruments used by the scholars of the Italian clergy in meteorology and seismology. This display had a clear apologetic purpose: it was meant to show that the Italian clergy were not as backward in science as certain quarters would have led one to believe. The collection of instruments made up one of the most interesting sections of the great fair, and it met with the approval of the Pope, who was known for his interest in the development of science.

When the exposition ended, the gifts of artistic value found homes in the various parts of the Pontifical rooms. But the Pope had a particular interest in the scientific collection, and he wished to see it kept together and used for scientific research. Fr. Francesco Denza suggested that the instruments be arranged in Gregory XIII's Tower of the Winds, which had been abandoned for some time. There, where an earlier incarnation of the Vatican Observatory had been located up until 1821, once again the observations could be taken up which would bring new glory to that ancient place of study and research. This new institute, the Specola Vaticana (or, simply, the Specola) would depend on the Supreme Pontiff through the Secretariat of State with respect to the projects it was to pursue. For administrative matters, on the other hand, it reported to the Prefecture of the Sacred Apostolic Palaces.

The observatory was founded at a propitious moment. Leo XIII saw that not only would such an institution offer the opportunity for official recognition for the Church in a field of international scientific research, but as a Papal observatory it would also offer the Holy See a way to be recognized by other nations as an independent body, and not a part of Italy.

The Pope appointed Denza as director of the new institute, and appointed to the staff those who had helped him in preparing the exposition. Denza was able almost immediately to enrich the existing array of instruments with others coming from the estate of the Marquis of Montecuccoli, who had operated a private observatory at Modena. Among these instruments there were two valuable Merz refractors, one of aperture 10.2 cm on an equatorial mounting and the second of 10.6 cm aperture on an altazimuth mounting. In addition there was a Stark coudé meridian telescope and four precision pendulum clocks for the measurement of sidereal time, plus another clock donated by Riefler of Munich for the jubilee of the Holy Father.

A rotating dome of 3.5 meters with a slit opening of 58 cm was soon installed on the Tower of the Winds to house the small Merz equatorial. This was the first of four domes that would be erected within the next few years in the Vatican. Thus the observatory soon began to operate.

Denza and the *Carte du Ciel*

Denza had, in fact, followed with close attention the most recent efforts of the astronomers of his day. At that time a number of astronomers under the leadership of the director of the Paris Observatory, Admiral Mouchez, were preparing an international initiative to photograph and measure the whole sky in a uniform way. From this would come the first photographically-based atlas of the stars: the *Carte du Ciel* (Map of the Sky). Based on these photos, a map and a catalog of the stars would be produced.

The first meeting of these astronomers, assembled in Paris in 1887

Above: Born at Naples in 1834, Fr. Francesco Denza donned the robe of the Barnabites in 1850. From his close association with Secchi, he had a great enthusiasm for physics, meteorology, and astronomy. After his ordination to the priesthood in 1858, he taught science at the Carl Albert Royal College of the Barnabites at Moncalieri and was a private tutor to the young Duke of Aosta. Primarily a meteorologist, he founded what became the Italian Meteorology Society and organized a number of meteorology conferences (most notably in Paris in 1878 and Rome in 1879). He was a pioneer in the measurement of ozone in the lower atmosphere. Denza was also interested in terrestrial magnetism, making field measurements throughout Italy, and seismology, helping to organize the first meeting of Italian seismologists 1887. Denza also wrote a book of popular astronomy, The Harmony of the Heavens.

under the patronage of the Academy of Sciences, set up the basis for this grandiose project and spelled out the direction to be taken. In a second meeting in 1889, the exact details of the undertaking were established. First of all it was necessary to set up the individual zones of the sky and to assign them to observatories that were willing and able to participate in the huge project.

Denza had a sharp eye for recognizing this as an excellent opportunity for the Vatican Observatory, and he proposed to the Holy Father that they become part of this initiative. He argued that participation in this program was the most fitting opportunity that could be offered to the observatory at that time, so that right at the beginning it would have the recognition necessary to carry out in the most effective way possible the mission given to it by the Pope: to nourish to the maximum extent the dialogue between the Church and the world of science.

The proposal was accepted, and the Vatican was given its swath of the sky to map. Denza obtained one of the standard astrographic telescopes to be used by all participants, and ordered from Gautier a macro-micrometer for the measurements of the Catalog plates. Afterwards the French government, with Admiral Mouchez as intermediary, bestowed the honor of Knight of the Legion of Honor on Denza in recognition of his work in having the Vatican

Above: Cover of a small booklet with lectures by Giuseppe Santalena, delivered on the occasion of the dedication of the Vatican Observatory's astrographic telescope, newly acquired to produce the "New Photographic Map of the Sky."

Observatory involved in the astrographic research of the *Carte du Ciel*.

In the history of astronomy, this enterprise represented the first great example of international collaboration in an astronomical program that was both well defined by agreements and worked out beforehand. In fact eighteen observatories located in countries on all continents participated in the project; later on, four other observatories joined. Each of the participating observatories was assigned a strip or zone of the sky, lying between two parallels of declination of the celestial sphere, and two series of photographs were to be taken: a short exposure to catalog the stars down to 11th magnitude by their brightness and position; and a longer exposure to reach 14th magnitude. Enlargements of the photographs were to be printed for the map.

The Vatican Observatory was assigned the strip between the parallels of +55 and +64 deg. To cover this completely and with some overlap would require 1,040 plates (later reduced to 540), both for the Catalog and the Map.

In order to guarantee that one would be able to distinguish true stellar images from spurious ones due to imperfections on the photographs taken for the mapping project or on the prints, multiple exposures were taken on a single plate, at least for half of the plates. For the Map, it was decided

Above: The Vatican Observatory Heliograph, a telescope designed for photography of the Sun. Standing with it is the technician Carlo Diadori.
Top: The Vatican Observatory clocks, necessary for the accurate tracking of stars, were obtained from the estate of the Marquis of Montecuccoli, who had operated a private observatory at Modena.

that three 40-minute exposures would be taken on each plate by closing the shutter and moving the telescope slightly in such a way as to place the three images at the vertices of a very small equilateral triangle at whose center lay the exact position of the star. Also, for the Catalog, it was decided

to take three exposures of 6 minutes, 3 minutes, and 25 seconds by moving the equatorial telescope between exposures by a small angle in declination. The exposures were much shorter than required for the Map, since the Catalog only went to 11th magnitude.

On every plate, whether for the Map or for the Catalog, before the exposure a faint reticle, which consisted of 26 x 26 little squares, one-half cm on a side, was photographed to serve as a reference frame for the precise determination of the position of each star. The triple image of each star was also useful for identifying those stars whose images might be covered by one of the lines of the reticle.

Now it became necessary to find a place for the *Carte du Ciel* telescope. This construction, close to the present-day location of the Vatican heliport and known today as the Torre S. Giovanni (St. John's Tower), has walls a good four and a half meters thick and is one of the few bastions still standing of the fortification that was called "leonina" because St. Leo IV had it built in 840 as a defense against the Saracen invasions. The circumference was large enough for the installation of an eight meter dome, and the strong vaulted arches guaranteed that the astrograph would be free of oscillations. Since the tower rose to a height of 20 meters above ground level on the highest point of the Vatican Hill, at about 100 meters above sea level, there was an unobstructed view on all sides. (And, of course, in those days electric lights from the city were not yet a serious problem for astronomers.)

The photographic equatorial telescope and the rotating eight-meter

In 1893, the observatory's equipment was still further enriched with a heliograph, a telescope especially designed for photographing the Sun. It had been made in Paris, the optics by the brothers Henry and the mechanics by Gautier according to plans of Prof. Janssens of Meudon. It had an objective of 14 cm diameter and a focal length of 2.15 m. With its magnification system, it reached 4.4 meters in length and produced on the photographic plate an image of the Sun 27.5 cm in diameter. The first of its kind in Italy, it was placed on the terrace of the monumental New Wing of the Museum of Pius VII, and for esthetic reasons it was covered by a sliding flat roof and not by a dome. Later on, it was placed under a small dome on the terrace of what is today the convent, Mater Ecclesiae, which John Paul II brought to the Vatican.

From the time he first came to

dome, which was also constructed at Paris in the Gilon workshops, were set in place on the Leonine Tower in 1891. The characteristics of the objective gave, on a photographic plate 16 cm square, a useful field 13 cm square, equivalent to two degrees on the sky, about four times the diameter of the Moon. The scale, therefore, was about 1 arc-minute/mm.

The Paris agreements provided that all of the 18 observatories participating in the *Carte du Ciel* would obtain instruments — not necessarily from Gautier and Henry — that had the same characteristics so that a perfect homogeneity of the results would be assured. The proven ability of the brothers Paul and Prosper Henry of the Paris Observatory in the fabrication of objectives particularly suited for sky photography was a decisive factor in convincing their director, Admiral Mouchez, to undertake the international sky-mapping project. Thus in 1891, with the sound placement of this principal instrument in order to collaborate in the great program initiated in Paris, the new Vatican Observatory had all of a sudden gone from its modest beginnings to become an important institute on the international scene.

On March 14, 1891, Leo XIII solemnly confirmed, with his Motu Proprio, *Ut Mysticam*, the foundation of the Specola Vaticana (see pp. 186ff); and with a financial contribution he assured its operation.

Above: Fr. Giuseppe Lais was born in Rome on April 15, 1845, attended the Roman College, and as a young man he had the good fortune of becoming known and appreciated by Secchi. In 1871, he entered the Congregation of the Oratorians founded by St. Philip Neri, and he was ordained a priest two years later. As Secchi's assistant, he acquired a facility with scientific instruments and a breadth of knowledge in astronomy unusual for his age. While at the Vatican Observatory, he would often invite young people to see the marvels of the sky from the terrace of his residence, and on summer evenings they would count shooting stars. One of those young people was Eugenio Pacelli, the future Pope Pius XII.

the Vatican Observatory as director, Denza took care of the continual flow of correspondence with other institutes and individuals. The remainder of his time he spent in editing the observatory publications. In 1892, he was appointed president of the Pontifical Academy of the *Nuovi Lincei* (PANL), the Vatican's academy of sciences at that time, but unfortunately he was not able to enjoy very long the fruits of his incessant labors to give new life to the almost superhuman task which he had taken up. But on December 13, 1894, following a Papal audience and a visit to the Cardinal Secretary of State, he suffered another stroke, and the following day he passed away. He was mourned by his colleagues throughout the world. He dedicated the last energies he had to the completion, in an incredibly short time, of the great work, and he made sure that it was provided for in the future.

Lais and the Photographic Project

Although Denza's death was a serious loss to the Vatican Observatory, he left the conduct of the principal project, the photographic map of the heavens, in competent hands, and thanks to the conscientiousness and tenacity of the vice-director, Fr. Giuseppe Lais, there was no delay in the project.

When Denza invited the Italian

Above: Lais working at the Carte du Ciel telescope.

to the observatory. His energies were depleted and coming to an end. In 1886, he had already suffered a cerebral hemorrhage, which paralyzed his right side. Fortunately, there was no damage to his intellectual faculties, so that, after learning to write with his left hand, he continued to dedicate himself

clergy to join in the collection of instruments for the exposition, Lais not only made a contribution of instruments, but he gave himself enthusiastically to organizing and preparing it. It is no surprise, therefore, that after the celebrations he was called by Leo XIII to assist at Denza's side, as vice-director,

December 26, 1921. Only a few days before, Pope Benedict XV had sent him a gold medal with a letter of thanks and praise for his constant zeal, exemplary accomplishments, and completely selfless dedication to serving the Holy See.

With youthful enthusiasm, Lais, already 45 years old, put himself to the job of carrying out the photographic program taken on by the Vatican Observatory at the Paris meeting. He went to Paris, and at the observatory of the French capital he learned the secrets of astronomical photography and of how to measure stellar positions on photographic plates.

The photographs for the Catalog began in 1894 and those for the Map in 1900. In the first years, he worked only every other night in the Leonine Tower, in close collaboration with Engineer Mannucci. However, when Mannucci was called to another job, it was left to Lais alone to carry out this

in the reconstruction of the Vatican Observatory.

Since he was dedicated to celestial photography, for many years the principal research work of the observatory, it is impossible to speak of this work without recalling the painstaking work which Lais performed for the Astrographic Catalog and the Map of the Heavens without interruption right until the time of his death. He showed no self-interest and completely dedicated himself to serving the observatory in this work. In fact, in addition to the fact that he received no payment for his work, on not a few occasions he made contributions to the photographic work, and he obtained grants from the Académie des Sciences. He also donated some costly and important magnetometers to the observatory.

Lais was a member of the PANL from 1875 onwards, and he was dean for many years. He demonstrated competency in a vast array of professional interests. His innumerable articles, listed in the *Proceedings* and the *Memorie of the PANL* and in the publications of the Vatican Observatory, range from meteorology to astronomy, from the history of science to archaeology, from chemistry to physics, from geophysics to celestial photography.

At the death of Fr. Bertelli in 1905, he was named to succeed him as president of the PANL. He fulfilled seven two-year terms of office and served almost to the time of his death,

Above: *Fr. Johann Georg Hagen was born in Austria on March 6, 1847. Having attended the Jesuit college at Feldkirch, Johann entered the novitiate in 1863 and was ordained a priest in 1878. He was then sent to the United States, where he taught mathematics for eight years at Sacred Heart College at Prairie du Chien, Wisconsin. He built a small observatory there and began to study variable stars. In 1888, Hagen was appointed director of the Jesuits' Georgetown College Observatory in Washington, where he published the first three volumes of his monumental work* Atlas Stellarum Variabilium. *During his stay in America, Hagen had become an American citizen. Fr. Hagen was also a deeply spiritual man. He served as spiritual director to Bl. Elizabeth Hesselblad, the Swedish/American woman who converted to Catholicism under his direction and founded the Swedish Branch of the Sisters of Our Savior, also called the Brigittines.*

monotonous work for 20 long years, a work which he accomplished most conscientiously and without counting the personal costs.

There is little doubt that the real reason why the Vatican Observatory did not have to face any particular difficulties, as did many other of the participating observatories, especially at the beginning, in the *Carte du Ciel* program, was that a person like Lais was right from the start able to dedicate himself totally to this work.

The "trials and strain" did not hinder him from enjoying a degree of contentment in certain by-products of the work itself as when, for instance, upon examining a plate taken on October 28, 1900, in the constellation of the Pleiades, he discovered an asteroid.

Of the 1,040 plates that the Vatican Observatory was to take for the Catalog, almost all were completed at the time of his death in 1921; and of the 540 plates for the mapping project, each requiring a two-hour exposure, 277 were completed and some of them enlarged and published. But the Catalog, consisting of a list of the stars by magnitude and position, was not yet worked out, even though this was really the principal goal which the observatory had set out to accomplish. As we shall see, it was left to Fr. Hagen to take up this work and bring it to a happy ending.

A Jesuit Comes to the Specola

Upon the death of Fr. Denza, it was difficult to find a new director. For four years, the Vatican Observatory continued its work under the administration of the vice-director, Fr. Lais; but Lais was completely taken up with photographing the sky at the telescope. Meanwhile, other scientists assigned to the observatory included a seismologist (following a strong earthquake felt in Rome

on November 1, 1895) and a meteorologist, who was named its director. Unfortunately, their attempt to help out with the astronomical work was less than successful, and the scientific reputation of the observatory began to suffer. Finally, in November of 1904, Pope Pius X appointed the archbishop of Pisa, Pietro Maffi, to reorganize the observatory and search for a new director. After more than a year of very delicate negotiations, in February 1906 the decision was finally made: the new director would be the Jesuit priest Johann Hagen (born in Austria, but by then an American citizen and director of the Georgetown Observatory in Washington). And, at the insistence of the new director, its work would concentrate solely on astronomy.

At the age of almost 60 years, the new director did not face an easy task. Helped in his first four years by Fr. Johan Stein, a young Dutch Jesuit astronomer and his future successor, Hagen put himself to the task of reactivating the observatory with youthful vigor and an iron will, reorganizing its astronomical and scientific programs.

For some time the astronomers had faced a serious inconvenience due to the fact that the space assigned to the Vatican Observatory was too spread out and that the employees did not have sufficient living and work areas. At the suggestion of Msgr. Maffi, Pius X, in 1906, sought to remedy the situation, and he very graciously put at the disposal of his astronomers his personal villa, today the headquarters of the technical division of Vatican Radio. (Leo XIII had had the villa constructed as a summer retreat in the shadows of the ancient Leonine fortifications; in his last years, the old Pontiff had stopped going there, and his successor, Pius X, had no intention of making further use of it.) The availability of space and the favorable position of this building, in direct contact with the second large

Sp. Vat. - Cometa Halley - 25 Magg. 1910

In 1910, on the occasion of the imminent passage of Halley's Comet, Fr. Johan Stein published an article about the famous legend that Pope Callistus III had anathematized the comet at its 1456 apparition in order to chase away all the calamities that it was said to bring. Stein researched this and found an authentic Bull of that time in which Callistus III, to implore God's help against the Turks, ordered solemn processions and prayers at the noontime sound of the bells. The near coincidence in time of the promulgation of the Bull and the comet's apparition, and the fact that the comet was visible during the first processions at Rome, led Stein to conclude that, since astrologers attributed to comets all kinds of dire happenings to humanity, the strange legend had its origin there.

tower still remaining of the fortification, made it the most appropriate for the observatory offices.

The massive tower next to the little villa was just as stable as the other tower that already supported the photographic telescope, and actually had a larger circumference. It was imme-

between the two large towers and lying next to the grotto, a copy of the one at Lourdes. On a lower floor of the same building, a small 10.6-cm Merz refractor on an altazimuth mount was placed for observations near sunrise and sunset of comets close to the sun. But the heliograph, which was at one time on top of the Museum of Pius VII, was placed on the terrace of the police barracks, which today is the monastery of the cloistered nuns, Mater Ecclesiae.

Access from the little villa to this terrace was provided by a passageway on the fortification wall that connected the two buildings. Now that the aesthetic considerations (which, on the *Braccio Nuovo*, had required that there be only a flat sliding roof to cover the heliograph) were no longer applicable, it was possible to cover this instrument with a 6.2-meter rotating dome, the fourth and last of the series.

In the meridian room on the top-

diately chosen for the large refractor, which was placed there in 1909. The mechanical structure was made by Gautier of Paris and the optics by Merz of Munich. It had an objective 40 cm in diameter, a focal length of 5.5 m, and it was placed under a dome 8.8 meters in diameter.

Under the dome and the strong vault that supported the telescope, there was a circular room in which Leo XIII had used to give audiences during the summer months when he resided there. For the semispherical vault, the Pope had commissioned the noted painter Ludwig Seitz to paint the starry sky over Rome at the time of the culmination of the constellation of Leo (an obvious allusion to the Pontiff's name) so that this constellation stood out among the others at the center of the vault. It was the best place imaginable to set up a museum. In fact, right in the center, a display case — at first a rectangular one and then an octagonal one — was placed, and the meteorites that had been donated to Pius X by the Marquis de Mauroy (see p. 149) were arranged in it. Around the circular wall there were cabinets where illuminated transparencies showing selected celestial objects were mounted.

In the meantime, in 1907, the small 10.2-cm Merz refractor with its 3.5-meter dome, which at one time was on the Tower of the Winds, was set up on the so-called half-tower — named for its semicircular form-intermediate

most floor of the little villa, the meridian telescope, or the transit instrument for measuring sidereal time, was placed. This was the Starke coudé that Denza had acquired from the equipment of the observatory of the Marquis of Montecuccoli. Later on, this instrument went out of use when a radio receiver for time signals to control the clocks was installed.

In the rooms on the lower floors, there was a clock room where there were four precision pendulum clocks with mercury compensators, three of them for sidereal time and one for mean solar time. These were also from the observatory of the Marquis di Montecuccoli.

A special room was set aside for the Gautier macromicrometer, which was used for measuring stellar positions on the plates of the Astrographic Catalog. Repsold micrometers, more precise and easier to handle, soon took the place of this instrument. Other rooms were assigned to meteorological instruments and to the library.

However, the arrangements that had been made for the observatory had one defect. In 1854, the wall that connected the two principal towers and provided a passageway between them had collapsed, so that to go from one telescope to the other one had to suffer the inconvenience of going down one tower and up the other. W. P. Doerr, an architect from Chicago visiting the observatory in the summer of 1906, suggested to Hagen that the problem could be easily resolved by means of a bridge, whose cost would be about $5,000. No sooner said than done. In a few months, Hagen raised $6,000 among his friends in America, and in 1907 he had his bridge. It was made of iron, 85 meters long, and held up by four scaffolded steel pillars. It linked the dome of the astrograph to the office complex, where soon the visual telescope and dome would be erected. And so the whole length of about half a kilometer of the old fortification wall with all of its three towers was made

Left: St. Peter's in 1924, before the Vatican Observatory moved to Castel Gandolfo. Note the two telescope domes on the wall behind the dome of St. Peter's, with the "American Bridge" connecting them across the part of the wall that had collapsed in 1854.

available to the Vatican Observatory. In the new space there was even room enough to place other instruments that were still in the Tower of the Winds; after 20 years of service to the observatory, that tower could finally and definitively be emptied out.

With like enthusiasm Hagen put his talents to the job of ordering and enriching the observatory library. He arranged it in the room that had once been the study of Leo XIII and in the adjoining room. He attempted to complete the series of publications and journals that were incomplete, either by buying them or requesting help from various authors and observatories; and in a short time, thanks to the kindness of his colleagues, he succeeded. With the Pope's approval, the ancient treasures of the Vatican Library that were of astronomical interest were also transferred to the Vatican Observatory. Among these were the complete series of the publications *Comptes Rendus* of Paris and the *Philosophical Transactions* of London.

On November 17, 1910, Pius X granted a special audience to the staff of the observatory to officially celebrate the new headquarters. The following year, as a commemoration of the eighth year of the pontificate of Pius X, the historical medal, which was customarily coined each year in gold, silver, and bronze and distributed to the members of the Papal Court and Ecclesiastical Dignitaries on the Feast of the Apostles

Peter and Paul, had inscribed on one side the allegorical figure of Astronomy speaking the words: *Ampliorem. in. Hortis. Vat. Mihi. Sedem. Adornavit* ("He has prepared for me a more ample seat in the Vatican Gardens"). Today, near the entrance to the chapel of Vatican Radio, in the little villa of Leo XIII, one can still see a plaque recalling the new housing of the Vatican Observatory.

From 1910 on, after his assistant, Stein, had been summoned back to Holland, Hagen was able to hire, first for four hours each day and afterwards for six, a secretary in the person of Pio Emanuelli, who, as an astronomy student, was able to help with calculations and with the correction of proofs.

Fr. Hagen's first worry was to organize the research for the Astrographic Catalog, whose timely completion was a question of honor. The Gautier macromicrometer proved to be too slow for such a large program, so Hagen decided to visit the various observatories that were already well along with the plate measurements and, based on their experience, he studied the best measuring technique and prepared for the best instruments to accomplish the task. So he had the firm Repsold of Hamburg make two measuring machines and, following the example

of the Paris observatory, he went to the nearby Institute of the Child Mary of St. Bartolomea Capitanio, and they kindly made three sisters available from 1910 until the completion of the measurements in 1921. The position and magnitude of the stars were taken two times with the plate in inverse positions by measuring respectively the coordinates of the image on the three-minute exposure and the diameter on the six-minute one.

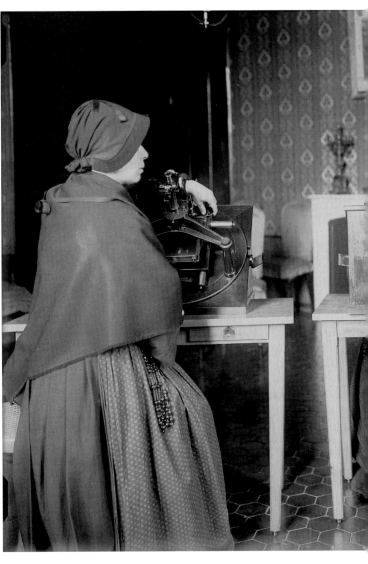

After years of tiring and difficult work, the mighty task was completed. It listed the brightness and the positions in rectangular coordinates of 481,215 stars. The 10 volumes were printed with extreme care by the Vatican Press, and the job, begun in 1914, was finished in 1928.

The Vatican Observatory was the fifth among the 18 observatories that had taken responsibility for the mapping to finish its section, preceded only by Greenwich, Oxford, Algiers, and the Cape of Good Hope. This was, indeed, an eloquent indication of the enormous difficulty of the undertaking, whose immensity could not be evaluated at the time of the initial enthusiasm. Hagen's selfless contribution to the Astrographic Catalog is

Above: In order to process the enormous amount of data produced in the Carte du Ciel program, Fr. Hagen arranged for sisters from the local Institute of the Child Mary to serve as "computers," measuring the star positions on the photographic plates.

even appreciated more if one realizes that he by personal choice was, and wished to remain, a visual observer of the old school. He never succeeded in developing an enthusiasm for astrophotography, which had already progressed so much since he was a young man, although later in his old age he did not disdain to use photographs for drawing the star charts for his *Atlas Stellarum Variabilium*.

Hagen did not have the same success in his second big project, the publication of the maps of the Vatican zones for which the observatory had assumed responsibility within the International Committee for the Photographic Sky Map.

It is true that Lais, thanks to substantial support from the French Académie des Sciences and from some French families, had already completed more than half of the required photographs; but when Hagen died in 1930, only 107 of the 540 maps were published. One of the principal obstacles was the large sum of money required for such a project, so much so that, precisely for that reason, some participating observatories right from the beginning had to withdraw from any publication of the charts. It was only after the Vatican Observatory was transferred (along with the astrograph) to Castel Gandolfo under Hagen's successor that the partly finished work was taken up again.

The best time for observing was during the war of 1915-1918, when the city of Rome was totally dark at night. But afterwards, because of the increasing nighttime illumination in certain parts of Rome, Hagen was forced to limit his observations to the northern parts of the sky. These serious problems, confronted in the attempt to complete the survey study of dark clouds were, as we shall see, the first reason for beginning to consider having a branch of the Vatican Observatory in a place more protected from artificial sky illumination.

The Rotation of the Earth

Besides his interest in the strictly astronomical problems, Hagen's untiring passion for scholarship led him to become involved in problems in physics, and in particular in proofs from mechanics for the Earth's rotation. In his work *La rotation de la terre et ses preuves mécaniques anciennes et nouvelles*, he gathered together and examined all of the attempts made up until that time to measure the Earth's rotation, and then he described two experiments of his own. The first was the application of a principle of mechanics whereby in a system of two masses rotating about a common center the rotation accelerates or decelerates as the masses change their distance from one another. Hagen clearly demonstrated this fact with an instrument that has become a classic. He called it the *Isotomeografo*, and he installed it on the basement floor of the astrograph tower. An 8.5-meter horizontal beam was hung by a two-strand steel cord about 7 meters long. Two small lead carriages weighing 90 kilos each could move on symmetrical tracks along the beam from the middle towards the ends and vice-versa. During the simultaneous movement of the carriages, the Earth's rotation caused the beam to undergo a small rotation in the horizontal plane, the rotation in the northern hemisphere being clockwise when the carriages moved from the middle to the ends of the beam and counterclockwise, just like the local component of the Earth's rotation, when they moved towards the middle from the ends. The principle had been known for a long time, but there was a long road to travel before the experiment, which showed quantitatively the effect predicted by the theory, was carried out.

In relationship to the proofs for the Earth's rotation, Hagen also repeated the experiment of the deviation to the east of falling bodies. This time, however, and for the first time in history, at the suggestion of Engineer Mannucci, he made use of a slow fall by employing an Atwood machine 23 meters high, which he installed in the Vatican Museum, within the triangular stairwell of Bramante. The experiment gave the predicted deviation to one percent. The principle that explains this phenomenon is also a simple one: when a body is lifted above the ground, since it is further from the Earth's axis than the ground itself, it is displaced a bit more rapidly than the ground as the Earth rotates. So Hagen won further acclaim as he was the first to make use of the Atwood machine, known since 1784, to show the rotation of the Earth. He received resounding applause from the participants at the fifth international congress of mathematicians at Cambridge in 1912 as they heard him give a paper describing the experiment.

After this, Hagen took up research on a second phenomenon, the "apsidal rotation," as it is called. It is connected with the free pendulum oscillation; as the amplitude of the oscillation decreases, the "oval spiral" made by the pendulum slowly gets "tighter towards the center" at the same time it goes on getting wider in the direction perpendicular to the "getting tighter" direction. This effect had been seen by many scholars, and it was the 19th-century Jesuit physicist Fr. Angelo

Secchi who was the first to intuit that it might be due to a "force of projection" acting on the pendulum in a direction perpendicular to the plane of oscillation, due to the Earth's rotation. Hagen interpreted the increase in the minor axis of the "oval spiral" as the major axis decreased as due to the law of areas, and he held that the phenomenon was a new proof of the Earth's rotation, completely independent of the one given by Foucault of the substantial stability of the plane of oscillation of the pendulum. This conclusion received further confirmation when one of his fellow Jesuits, Fr. E. F. Pigot, director of the Seismological Observatory of Riverview College in Sydney, Australia, was asked to repeat the experiment in the southern hemisphere; he verified that the apsidal rotation occurred, as was to be expected, in the opposite sense to that observed in the northern hemisphere.

And so Fr. Hagen, 83 years old before death took the pen from his tireless hand on September 5, 1930, could look back contentedly at a successful life full of work to the good of science and the Church. Especially linked to his name will always be the reorganization of the Vatican Observatory. A crater on the Moon has been named for him.

The Move to Castel Gandolfo

After the death of Fr. Hagen, Pope Pius XI called upon another Jesuit, Fr. Johan Willem Stein, to succeed him. Stein was not new to the Vatican Observatory, since he had been an assistant to Hagen from 1906 to 1910.

Above: Proof of the Earth's rotation was derived from Fr. Hagen's "Isotomeografo," which detected the tiny Coriolis force acting on weights moving in a North-South direction. When the two weights on wheels, each massing 90 kg, moved towards the center of the device, the change in their position relative to Earth's spin would result in a slight counterclockwise motion of the beam on which they moved, as indicated by the arrow painted on the wall.

The first challenge facing the new director was to find a new location for the observatory.

Writing in 1932, Stein described the move to Castel Gandolfo by noting the problems of encroaching city lights on the telescopes at the Vatican. He then wrote:

To keep up with the times the Specola needed two things: a new mounting for the visual telescope, and a new high-power astrograph with the necessary auxiliary instrumentation. But who would have taken on the responsibility of the costs, when the conditions of the Specola were so precarious? And furthermore where was there a location suitable for a new astrograph, since the two principal towers were already being used?

The solution was found by His Holiness, a solution which was acclaimed by astronomers in Italy and abroad. His Holiness would make available his Pontifical Villa at Castel Gandolfo, 22 kilometers from Rome, on the condition that the surroundings and the atmosphere should prove to be suitable for a branch of the Specola.

In order to verify this condition, on some nights during November and December 1931 we photographed in the villa some star trails and pairs of double stars at various altitude and azimuth positions in the sky in order to sample the sharpness and clarity of the images. The result was fully satisfactory. On some nights the view of the Milky Way was truly beautiful and enchanting. To the East the atmosphere was usually clear right to the horizon. It is true that quite low to the Southwest over the Tyrrhenian Sea there normally hung a stretch of clouds, while far away on the Northeast horizon was seen the glow of the Eternal City; but one never, or at least seldom, observes so close to the horizon.

Following the report of our experiments, it was in principle decided by His Holiness that the branch of the Specola would be established and a new first-class photographic telescope acquired.

There remained the problem of selecting the most suitable place for the new branch. At the beginning His Holiness had called our attention to land in the huge gardens of the Villa Barberini, joined to Castel Gandolfo. But quite soon his idea changed and he showed a preference to see the new telescope mounted on the top terrace of the Castle, unless there would arise problems of a technical nature. In fact, that terrace from a height of 430 meters above sea level provides a magnificent view and seemed to be an ideal place to set up an observatory.

The plans for the new Vatican Observatory were prepared by the famous firm Carl Zeiss of Jena. Prof. Paul Guthnick, director of the Berlin Observatory at Neubabelsberg, took a great deal of interest in the new undertaking and made his vast experience available to Stein. The observatory was to be equipped with the most powerful instruments, not necessarily the most massive, but of a high quality and practicality, matching that of the best modern observatories.

Construction began in 1932, and by 1935 it was mostly complete. The new visual refractor was installed under a large wooden dome, 8.5 meters in diameter, resting on the massive round construction of the palace's ancient

Above: *Fr. Johan Willem Stein was born at Grave in the Netherlands on February 27, 1871, entered the Society of Jesus in 1888, and was ordained in 1903. He studied physics and astronomy at the University of Leiden with H. A. Lorentz. From 1906 to 1910, he was assistant to Hagen at the Vatican Observatory. The most important of Stein's strictly astronomical works was his completion in 1924 of Hagen's great work on variable stars,* Die Veränderlichen Sterne. *He continued to observe until, at 78 years old, he could no longer manage the telescope; he then dedicated himself to historical research. He was a counselor of the Italian Astronomical Society and a Corresponding Member of the Royal Dutch Academy of Sciences. Queen Juliana made him a Knight of the Lion of Holland, and a lunar crater is named for him.*

spiral staircase. Zeiss had previously planned that this telescope would follow the measurements of the old Rome refractor. It was to have been used for observing Hagen's dark nebulae at the planned observing station in the southern hemisphere. Now, since a new mounting and dome would be required, it was not worth the trouble to bring this old instrument to Castel Gandolfo. Nevertheless, it was planned to use the old lens, but, when a careful optical test was performed, it showed such great stress that the firm suggested the purchase of a new lens, with the assurance that they would take back the old lens and compensate for it in a small way.

Thus Zeiss provided a completely new telescope with an equatorial mounting, a 40 cm objective of 6 m focal length, together with a set of nine eyepieces and various accessories. The instrument was also provided with a Graff photometer for observing variable stars and with a micrometer for measuring double stars. Later on, a Danjon stellar interferometer was added. This was constructed in the observatory workshop and could be used for the determination of distances and positions of double stars and the diameters of planets.

In the second rotating dome, 8 meters in diameter, also made of wood constructed on the solid foundation offered by the northeast corner of the palace, the principal instrument of

the observatory was placed: the Zeiss Double Astrograph. It consisted of a refractor with a 40 cm four-lens objective of 240 cm focal length and a reflecting telescope with a 60 cm parabolic mirror of 200 cm Newtonian focal length and an equivalent 8.2 m focal length at Cassegrain focus. Both instruments plus two finders and a guide telescope were rigidly linked together and mounted on the same polar axis.

A large spectrograph could be mounted on the reflector for astrophysical research. The four-lens astrograph allowed 30x30 cm photographs with image correction to be taken; it was particularly suitable for photographic observations of variable stars and for the photographic determination of the positions of minor planets and comets. Two large (61.2 cm diameter) flint prisms of refracting angle 4 and 8 degrees respectively could be attached singly or together at the upper end of the reflector or refractor, thereby allowing spectra to be taken over large fields.

For the various measurements to be carried out on the photographic plates, Zeiss supplied a large Komess coordinate measuring machine, a Hartmann spectrocomparator, and an eclipse comparator, which was especially useful for variable star research and for minor planets. For the measurement of photographic stellar magnitudes, the firm Ascania of Berlin supplied a microphotometer, which

Top: The new location for the Vatican Observatory, on the roof of the Papal Palace in Castel Gandolfo, provided, in the words of the director Fr. Stein, "a magnificent view ... an ideal place to set up an observatory."

ted with a new invention of Zeiss, a tribune that could be raised, lowered, and rotated.

The offices, the map collection, the archives, the Starke meridian telescope, and the storage of batteries for direct current supply were arranged on the top floor, the fifth, of the Apostolic Palace. A large room was reserved for the operations center for keeping sidereal time. A Riefler pendulum clock of great precision served as the driver for four other clocks, two of them placed in the domes. Controls were effected by a radio receiver tuned to time signals from Paris. The library was arranged on the fourth floor, while the collection of meteorites and minerals were placed in a room on the second floor.

The Astrophysical Laboratory

The idea to associate an astrophysical laboratory specializing in spectroscopy with the astronomical observatory was motivated at least in part by the rich collection of meteorites in the Vatican Observatory's possession. But certainly it was well understood that it is almost exclusively through spectroscopy (see p. 88) that precious information can be obtained about the age and structure of the objects from which the meteorites came. Spectroscopy has, in fact, the advantage that it requires only a very small amount of this precious material for the investigation and that the concentration of the scientifically interesting elements is quite often very small. At Stein's suggestion, Pius XI appointed Fr. Alois Gatterer to draw up a plan for an astrophysical laboratory, which he would direct.

The Pope provided space on the ground floor of the Papal Palace for the laboratory, and in the summer of 1933 work began on setting it up. Almost all work was completed by the following year. The laboratory was equipped with a powerful array of spectrographs, providing all that was necessary for this difficult field of research. At the beginning there was a Zeiss grating spectrograph with a photographic camera and a large GH spectrograph of the firm Steinheil of Munich with three prisms for the study of visual light and two quartz prisms for the ultraviolet. In order to make full use of the resolving power of the prisms, Gatterer had a

allowed measures according to the Hartmann method (subjective) or the thermoelectric method (objective). Later on, this instrument was improved with the addition of a photomultiplier tube. In order to transport the observer to a convenient observing position at the telescope, both domes were outfit-

Above: The Zeiss Double Astrograph, one of only four ever made, consisted of two telescopes on a common mount; they would be able to take spectra of stars of the same part of the sky simultaneously with photographic plates sensitive to two different ranges of wavelengths. This was an important ability in the 1930s, before panchromatic photographic film (sensitive to all the colors of visible light) was available.

160 cm focal length camera made for the spectrograph, and it proved to be excellent; in fact, it was so excellent that the company was called upon to build such large spectrographs for other institutes. Little by little other spectrographs were added to the original ones. There were two especially for the ultraviolet: one with two prisms by the firm Halle of Berlin and an intermediate one of quartz by Zeiss. For the visual and infrared, a three-prism spectrograph was purchased from Zeiss. It could be used in an autocollimation mode so as to double the dispersion and was, therefore, especially useful in examining spectra of elements that had many closely packed lines. Later on, in 1957, through the generosity of Pope Pius XII, a grating spectrograph was purchased from the Jarrel-Ash Company of Boston, and a large spectrograph with three prisms and one half prism was built.

Various kinds of electrical current provided the workplaces with the energy necessary to excite to luminescence the materials under examination. For the study of the spectrographic plates, the laboratory had at hand the most modern instruments for measuring the wavelengths and the strength of the spectral lines. Later on, there was added to all of this equipment a large Zeiss photoelectric photometer, a projection comparator for comparing spectra on different plates, and an ingenuous universal stand for holding

meteorites to be examined with the spectroscope. These were all constructed, according to designs by Gatterer, by Br. Karl Treusch, the observatory's skilled mechanic.

The laboratory was also furnished with a sophisticated darkroom for the development, enlargement, and print-

Above: The Zeiss refractor telescope, installed in a dome over the main stairwell of the Papal Palace in Castel Gandolfo, has an aperture of 40 cm and a 6 m focal length, making it ideal for studying double stars, planets, and other objects requiring high magnification. At the eyepiece in this photograph is Fr. Sabino Maffeo, author of this chapter.

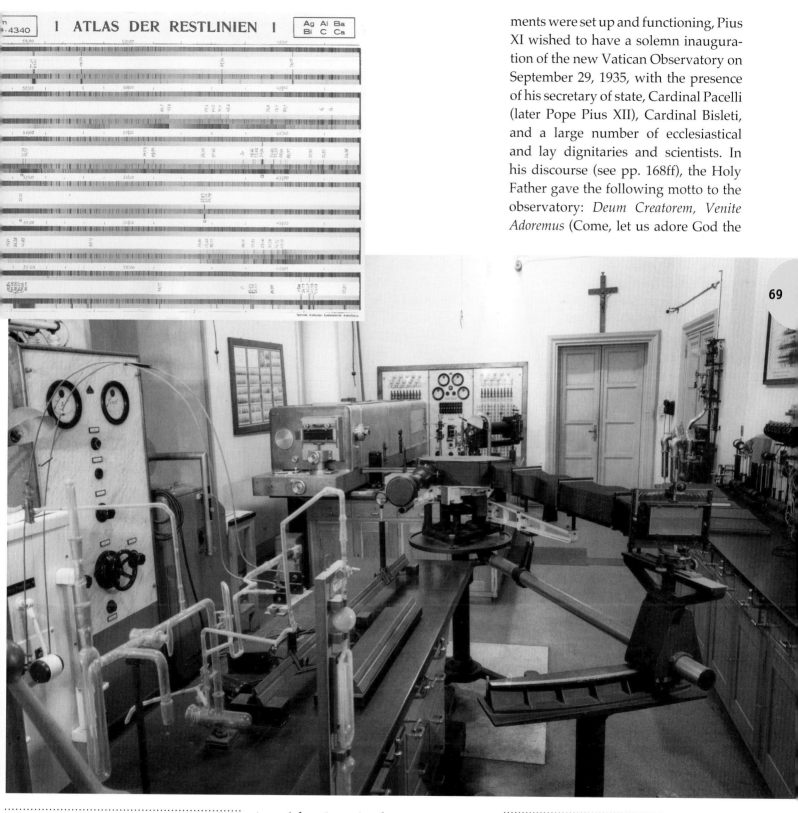

4340	I ATLAS DER RESTLINIEN I		Ag Al Ba
			Bi C Ca

ments were set up and functioning, Pius XI wished to have a solemn inauguration of the new Vatican Observatory on September 29, 1935, with the presence of his secretary of state, Cardinal Pacelli (later Pope Pius XII), Cardinal Bisleti, and a large number of ecclesiastical and lay dignitaries and scientists. In his discourse (see pp. 168ff), the Holy Father gave the following motto to the observatory: *Deum Creatorem, Venite Adoremus* (Come, let us adore God the

Top: *The work of the Spectra Lab led to the publication of several detailed atlases of spectral lines and eventually led to the foundation of the journal Spectrochimica Acta.*

ing of the plates for the spectral atlases and with a chemical section, small but well equipped, with its own supply of gas, vacuum, compressed air, and liquid air for the preparation of the substances to be examined.

After the construction work had been completed and most of the instru-

Above: *Upon the move to Castel Gandolfo in the 1930s, a state-of-the-art laboratory was set up to measure precisely the spectral lines of metals, which could then be compared against the spectral lines observed in stars.*

Creator). This invitation, inscribed in the bright marble that stands out on the south wall of the dome of the double astrograph, has been an incentive to the astronomers who for years have gone there to their night's work, and it is still an inspiration to the visitors who come to admire the work of a Pope who was an enthusiastic champion of the sciences.

The work of the Spectral Laboratory consisted principally in the production of spectral atlases. The first were the arc and spark spectra of iron, originally intended for in-house daily use in the laboratory; but they proved to be so useful that it was decided in 1935 to publish them. The first atlas consisted of 13 pages of figures and presented only the ultraviolet portion of the iron spark spectrum from 4650 to 2242 Å; the second with 21 plates of figures presented the arc spectrum also in the visible from 8388 to 2242 Å. A second edition, revised and updated, was published in 1947.

Having witnessed the success of the first atlases and how enthusiastically they were received by spectroscopic institutes, Gatterer and Junkes decided to compile a spectral atlas for each of the chemical elements that might be of interest for future spectral analyses. After 12 years of patient labor, this work was completed in 1949. Over the next 25 years, other atlases followed. The death of the last of the spectrographers, Fr. Salpeter, in January of 1976,

finally led superiors to the decision to close the laboratory.

At the time that the laboratory was founded, there did not exist a journal dedicated specifically to spectroscopy. To improve this situation, Gatterer, in 1939, began an international journal dedicated exclusively to spectroscopy, entitled *Spectrochimica Acta.* In 1944, after the publication of the second volume, the journal was forced to cease publication because of the war. When hostilities ended, the publishing house, resurrected from the ashes of war, considered starting the journal up again, and they asked Gatterer whether, for the difficult postwar period, it might be published in the Vatican. Pope Pius XII, who was always sensitive to whatever might nourish scientific collaboration among nations, not only gave his approval to this very important work in the field of spectrochemistry, but he also helped to ensure its future with some financial assistance. What came to be known as the Vatican Volume contained reports of spectroscopic research carried out during all of the war years in Great Britain, Belgium, France, and the Soviet Union. When the general situation returned to normal in 1949, it became possible to entrust the journal to the Pergamon Press. Since 1953,

when Gatterer, the founder of the journal, died, no member of the observatory has served on the editorial board. Towards the middle of the 1960s, the journal had reached an annual mean total of 2,000 pages, and it was, therefore, divided into two sections: section A for molecular spectroscopy and section B for atomic and analytical spectroscopy. From at least 1980 on, section B has been generally recognized as the most prestigious journal in the field.

SPECTROCHIMICA ACTA

COMMENTARIUM SCIENTIFICUM INTERNATIONALE
TRACTANS DE RE SPECTROCHIMICA

COLLABORANTIBUS

G. BALZ-STUTTGART, J. BARDET-PARIS, R. CASTRO-UGINE, G. H. DIEKE-BALTIMORE, Wa. GERLACH-MÜNCHEN, G. HANSEN-OBERKOCHEN, G. R. HARRISON-CAMBRIDGE, MASS., P. JOLIBOIS-PARIS, M. LOEUILLE-PARIS, J. M. LÓPEZ de AZCONA-MADRID, R. MANNKOPFF-GÖTTINGEN, O. MASI-MILANO, W. F. MEGGERS-WASHINGTON, D.C., K. PFEILSTICKER-STUTTGART, J.M. PHÉLINE-UGINE, R. A. SAWYER-ANN ARBOR, MICH., R. SCHMIDT-AMSTERDAM, B.F. SCRIBNER-WASHINGTON, D.C., D.M. SMITH-WEMBLEY, F. TWYMAN-LONDON, R. A. WOLFE-ANN ARBOR, MICH.

EDITUM A

R. BRECKPOT
LOUVAIN

A. GATTERER
CITTÀ DEL VATICANO

H. KAISER
DORTMUND

E. H. S. van SOMEREN
LONDON

L. W. STROCK
SARATOGA SPRINGS, N. Y.

VOLUMEN III.
EXORNATUM 176 FIGURIS TEXTUI INSERTIS

APUD
SPECULAM VATICANAM
ASSENTIENTE
SPRINGER-VERLAG, HEIDELBERG-GÖTTINGEN-BERLIN

CITTÀ DEL VATICANO
SPECOLA VATICANA
1947-1949

....

Above: *The "Vatican Volume" of* Spectrochimica Acta. *This journal, founded at the Vatican Observatory before the war, was actually edited and printed in the Vatican in the years following the war when resources for such publications were scarce throughout Europe.*

Castel Gandolfo before the War

With the refoundation of the Vatican Observatory at Castel Gandolfo, Pius XI also arranged that the running of it would be entrusted to the Society of Jesus. Thus, since that time, it has been the responsibility of the Society's Father General to propose to the Pope the person to direct the observatory and to provide for it an adequate number of Jesuit scientists. And so it was that Father General Ledókowski responded immediately to the request for collaborators which Stein and Gatterer had addressed to him, and he assigned a certain number of Jesuits to the staff. Thus, little by little, between 1933 and 1940, an international community began to form, consisting of eight fathers and four coadjutor brothers. Then as now this community is a unique example of a group of Jesuits dedicated to scientific research under the direct administration of the Holy See.

Pius XI assigned to his "sons of the sky," as he one time jokingly referred to his astronomers, a fine apartment on the second floor of the Pontifical Palace on the side facing the lake. Five fathers helped Stein in the strictly astronomical research, while Gatterer, director of the astrophysical laboratory, had a collaborator from outside in expectation of the arrival of Fr. Junkes.

Two of the four brothers were mechanics and took care of the maintenance of the equipment and the construction of new instruments. One of those two soon left the observatory; the other one, Br. Karl Treusch, stayed there until 1978. Before becoming a brother, he had gained invaluable experience as an employee of the famous firm Carl Zeiss of Jena, experts in telescope construction; thus he was destined from birth for the Vatican Observatory!

Right from the beginning it was a point of honor for the new Vatican Observatory to carry on the works taken up previously and to bring them as soon as possible to completion. The first such case was the publication of the eighth volume of the *Atlas of Variable Stars*, which Hagen had left unfinished. The maps were no longer drawn as before but rather produced in the observatory's laboratory by a photographic process; they were the first photographic maps of their kind that give a view of the sky as it really appears to the human eye. With this volume IX, in 1941, Hagen's great *Atlas of Variable Stars* was completed.

Another task that required completion was the continuation of the photographs for the *Carte du Ciel* and the publication of the respective charts. Lais had already finished 277 of the 540 plates required but, at the time of Hagen's death, only 107 maps had been printed. For that matter, the other observatories that were participating in this international effort were also in a similar situation. On the other hand, Hagen was too occupied in his last years with his own research and in the more important work of compiling the Astrographic Catalog. Because of this, the reproduction of Lais' plates had made very little progress, and furthermore, for lack of a competent observer, no more plates were taken after his death. When the main work of getting the new Vatican Observatory in order was finished, and enough help had been obtained to carry out the astronomical research, Stein put himself to the task of completing the *Carte du Ciel*.

In the meantime, the sky photography with the astrograph, still located in the Vatican, was not neglected. The number of plates required to complete the Vatican zones of the *Carte du Ciel* still had to be completed. But it was soon realized that this instrument should also be transported to Castel Gandolfo without further delay. Pius XI approved, and in 1938, during his last walk before returning to Rome from his summer stay at Castel Gandolfo, he looked around for a place in the gardens of the Villa Barberini where one might consider putting the new installation. But it was only in 1942 under his successor Pius XII, and despite the fact that the war had already begun, that it was possible to effect the transfer. On that occasion, the old refractor and the dome that housed it were carefully restored after 50 years of service. An epigraph in Latin, inscribed on the north telescope pier, recalls this new restoration that occurred 50 years after the refoundation of the Vatican Observatory.

The Specola and World War II

In the early years of the war, the work of the observatory (as a part of the neutral Vatican City State) was only slightly affected. All that changed after the Allied landing at Anzio on January 22, 1944; now the war was in the sight of the astronomers. The populations of Albano and Castel Gandolfo were forced to leave their houses, and the greater part of them took refuge either in the Papal Villas or in the Papal Palace, trusting that the fighting factions would respect the extraterritoriality, which was well indicated to aircraft by the Papal colors. About 2,000 persons found refuge in the palace for more than four months, 127 of them occupying 10 rooms of the astrophysical laboratory, all of them with equipment disassembled and in disarray and filled with baggage and belongings of every sort which the evacuees had brought with them. Even the space on the ground floor of the *Carte du Ciel* astrograph was occupied by a family from Albano.

Right from the beginning, the situation proved to be difficult and dangerous, because of the intensified air bombardment of the German positions whose command station was situated in some small villas overlooking Lake Albano, very close to the summer location of the College of the Propaganda Fide, part of the extraterritorial area of the Villa Barberini.

On the 24th of January, even though the palace was already full, the Vatican Observatory put up for some days 84 Jesuits who had to suddenly evacuate the novitiate house at Galloro in the neighboring town of Ariccia. We read in the diary written in the telegraphic style of Stein:

On the 28th of January the Superior exhorted the inhabitants of Castel Gandolfo to offer themselves willingly (to avoid worse fates), since 40 men were needed to work for the Germans. He joined in the work squads. The same happened the following day with a handful of fifteen men. A baby born in the Palace: Eugenio.

1 February. Albano bombarded: Dead about 15 cloistered nuns, 3 Christian brothers, 1 Giuseppino brother, some lay people. Our fathers and brothers helping.

2 February. 14 large caliber bombs fallen in the Villa Barberini (one not exploded).

10 February. Bombardment in the Villa Propaganda Fide. Many (c 500) dead or injured. ...Father Pignatelli, Treusch and Timmers to the Villa to help dig out the vic-

tims. Father Pignatelli the whole night. The aqueduct broken, lacking water.

12 February. Father Provincial agrees to let the sisters pass through our corridor. Cloister limited to our rooms downstairs.

On the night 13-14 February. A machine gun bullet has hit the photographic dome, has made a hole, then hit a small beam, and then fell unexploded on the floor.

18 February. Evening. Big pieces have fallen behind the Church of C.G. The burgomaster is dead. Windows of the Church broken and the windows of the Palace on the lake side, already three holes in the photographic dome. Father Pignatelli with four brothers and the donkey lead the 5 cows on foot from Galloro to Rome (to the Gesù).

Between the 16th and the 21st of

The Vatican Observatory and the war
After the invasion of Anzio in January 1944, the area around the Papal Palace and gardens were subjected to heavy bombardment.
Some 2,000 refugees from neighboring towns took refuge for most of that winter and spring in the Papal Gardens attached to the Pope's Summer Residence in Castel Gandolfo, which was neutral territory during the war.

February, given the increasingly dangerous situation, the Jesuit community, with the Holy Father's consent, moved to the Jesuit General Curia (the Jesuit headquarters) in Rome. Only Fr. Zirwes and Br. Treusch stayed behind to share the lot of the refugees and help alleviate their sufferings.

We quote again from the diary:

21 February ... Instruments sent to the Vatican, under the library.

30 May: Father Zirwes arrives in the Curia. Left only Brother Treusch at C.G.

Among the many things that Treusch did to help the refugees there was keeping the men occupied, putting them to the building of a cistern to help solve the chronic lack of water and to the digging of an

Opposite left: The Cryptoportico, the ruins of the stables from the palace of Roman Emperor Domitian built around A.D. 80, were once again pressed into service to shelter animals — and their human caretakers — during the winter of 1944.
Above left: Refugees doing their daily tasks next to the dome of the Carte du Ciel telescope.
Above right: A papal audience hall in the summer residence became sleeping quarters for many of the 2,000 refugees.
Top: On February 10, American fighter planes mistook the villa house for the College of the Propagation of the Faith, where refugees were staying, for a German military headquarters; 500 people were killed when it was destroyed. Members of the Vatican Observatory worked all night to help rescue survivors.

underground hideaway for protection from the bombardments. This difficult situation went on until the first days of June 1944 when, with the arrival of the Allied forces, the war front moved to the north of Rome. At that time all of the evacuees returned to their houses, most of which had been damaged or destroyed. The Jesuit Community returned to the Vatican Observatory on the 21st of that month.

On February 1, 1947, Stein notes in his diary: "Monsignor Montini communicates: the industrial complex and the items in storage of the Zeiss firm transferred to Russia as war-time reparations." And so, due to the war, the first objective of the visual telescope, which had been sent back to Zeiss in Jena to be corrected, and the first G 80 spectrograph designed by Gatterer were lost.

After the War

When the war ended, the astrograph was again put to work. Making use of every favorable night, by about the middle of 1953 Junkes and then De Kort managed to finish the number of plates that had to be taken to complete the *Carte du Ciel*. Then, by the end of 1955, Br. Mattieu Timmers, at first, and then, after his untimely death, Br. Luigi Puhl, under the direction of Fr. Peter Albert Zirwes, completed the enlargements and the production of the photostatic copies. And so, after 55 years from the beginning of work, the Vatican Observatory brought to completion the most laborious part of the task that had been assigned to it in that faraway year of 1889 by the Permanent Committee of the International Sky Mapping Program (*Carte du Ciel*). The Catalog had been completed in 1928. All stars down to the 14th magnitude in 10 zones of the sky between 55 and 64 degrees of declination had been reproduced on 540 charts, 26 x 26 cm square. If we lined them up, edge-to-edge we would have a 140-meter ribbon! About 100 copies of the whole work were printed, and about 90 of them were sent to the chief astronomical observatories in the world. The official announcement that the Vatican Observatory had completed its work was given at the General Assembly of the International Astronomical Union (IAU), in Moscow, in 1958.

In addition to the works described above, which we might call traditional for those times, the observatory, with its new high-quality equipment, set itself to another specific goal: research on the structure of our own stellar system in order to understand how stars are distributed in the Milky Way.

To know the stellar component of the Milky Way, one selects a certain number of stars as probes of given regions. For instance, the study of variable stars, specifically Cepheid and RR Lyrae stars, allows us to determine distances to selected regions and, therefore, to estimate the dimensions of the Galaxy. Spectral studies of the stars give us information on the composition, temperature, atmospheric pressure, magnetic fields, their motions, and hence how the stars are rotating with the Galaxy. The extinction and polarization of starlight gives us information on the kind of material existing in interstellar space and on the interstellar magnetic fields. The analysis of eclipsing variable stars allows us to determine the masses of the component stars. Multicolor photometry gives us information on variability, its causes, etc.

In order to complete the research in a limited amount of time, spectral classification was begun on stars in 15 regions near the galactic equator. The high quality of the spectrograph with the 4-degree objective prism allowed one to obtain classification spectra of stars as faint as 14th magnitude in 4 hours of exposure, a limit probably never reached elsewhere with a refractor.

Research in stellar spectroscopy had been successfully begun by Dr. Hermann Brück and by a young Hungarian Jesuit, Fr. Mátyás Tibor, before the war. Later on, in 1940, Junkes joined the staff, when Tibor had to leave the Vatican Observatory for health reasons, and by attaching an objective prism to the reflector, Junkes obtained a series of high-quality reference spectra of standard stars.

But these first projects also showed the limitations of the Zeiss double astrograph for taking stellar spectra. On the one hand, the achromatic correction of the refractor to accommodate the photographic region placed limits on the usable bandwidth for spectroscopy. On the other hand, the attachment of two prisms at the same time did not give good results for long exposures because of the instrumental flexure due to the excessive weight of the prisms. These are the reasons why, a few years later, a Schmidt telescope was purchased.

Right after the war, Fr. Walter Miller arrived at the Vatican Observatory from the United States. He pursued research on faint variables in selected Milky Way regions. In his nine years at the observatory he accumulated more than 3,500 photographic plates, almost all taken directly by himself, the remainder borrowed from other observatories. Using this precious collection of plates, he was able to classify about 500 new variables, which soon became known as Vatican Variables (VV). This work drew the attention of a good number of specialists who had a particular interest in the study of the structure of the Galaxy. During those same years, De Kort concentrated his research on RR Lyrae type variables and eclipsing variables.

This later field of research became a dedicated area of Fr. Daniel O'Connell, who succeeded Stein as director of the Vatican Observatory in 1952.

Above: Born in 1896 to Irish parents living in Great Britain, Fr. Daniel O'Connell entered the Irish province of the Society of Jesus in 1913 and studied astronomy in Dublin and at Harvard. He came as director to the Vatican Observatory in 1952 from Sydney, Australia, where he had been director of the Observatory of the Jesuits' Riverview College since 1938. When he arrived in Rome, he brought with him about 3,000 plates taken at that observatory for the study of variable stars in the southern Milky Way. One of the noteworthy results of his research was the discovery of the "O'Connell Effect," which refers to binary systems in which there is a rotation of the line of the apsides. These are very important systems because, from an analysis of the light curves, the mass and internal structure of the components can be determined.

The Green Flash

From 1954 to 1957, O'Connell, in collaboration with Br. Karl Treusch, carried out an interesting piece of research on the so-called green flash. This striking phenomenon occurs when the atmosphere is particularly clear and calm: under these conditions, the last segment of the Sun at sunset (or the first at sunrise) appears not red but green. For a long time, scientists were divided in their opinion as to whether this was a subjective phenomenon — that is, an optical illusion — or an objective reality. Spectroscopy and color photography had finally established that the phenomenon was objective, but no one had yet put himself to publishing color photographs of it. And so O'Connell put himself to the task. As he related it, the idea to do this came to him from the fact that, although he had tried many times in various places to observe the green flash, he first succeeded in seeing it, and very clearly, looking out over the Mediterranean from the window of his office at Castel Gandolfo. And so he conceived the idea of obtaining color photographs of the phenomenon by using the observatory's telescopes so that he might, if nothing else, confirm its objectivity.

The observatory's mechanic, Treusch, who by this time had also become an expert in telescopic photography, was rather skeptical at the beginning because he could see the enormous difficulties to be confronted. In addition to the fact that it is not easy to predict, the phenomenon lasts for an extremely short time, and it is, therefore, very difficult to photograph. But after the first tests, he became enthusiastic about the work — so much so that, as O'Connell himself says, the successful outcome was due principally to the talent and the patience of his capable assistant.

The results of this research were gathered together by the author in a splendid 200-page volume, *The Green Flash and Other Low Sun Phenomena*. It contains a selection of hundreds of color photographs of the green flash and of the Sun taken at sunrise and sunset, and it is the first publication of its kind (see pp. 22-23). Since the green flash has a certain poetic fascination for the public at large, it is without a doubt the research which is the most popularized of all the other programs of the observatory.

The Schmidt Astrograph

A new type of telescope called a Schmidt astrograph, invented in the last years before the Second World War, produced a veritable revolution in the field of Milky Way research. In 1949, Stein proposed the purchase of this type of telescope to the Holy Father and, cognizant of the need, Pius XII gave his unhesitating approval. And so, towards the end of that same year, the telescope was ordered from the firm of Hargreaves and Thomson of London. Five years later, work began on the construction of the Schmidt building in the gardens of the Villa Barberini. The new building was joined to the dome that had already been built in 1942 to house the *Carte du Ciel* astrograph. The telescope was delivered in 1957; a few days before his departure from Castel Gandolfo that summer, the Holy Father was kind enough to visit and bless the new installation and tele-

scope. It was strictly a private ceremony with only the fathers and brothers of the Vatican Observatory taking part (see pp. 174-175).

Several years were required for its installation and testing. It was only in 1962 that observational programs could begin with it.

The instrument is made of a 5-meter-long tube at the bottom of which is a spherical mirror 98 cm in diameter. The 65-cm corrector plate, which determines the aperture, is mounted at the center of curvature of the primary near the tube opening. The focal length is 2.4 meters. The 20 x 20 cm photographic plate is placed at the focal plane, which lies about halfway down the tube. The usable field is about 5 x 5 square degrees. This means that one side of the plate covers about 10 times the apparent diameter of the full Moon, a field more than six times larger than that of the *Carte du Ciel* astrograph.

The telescope was primarily planned for stellar spectroscopy. For this purpose, it is possible to mount a combination of three objective prisms at the tube opening; these prisms are among the most powerful in the world. Two of them, with angles of refraction of 4 and 8 degrees, respectively, already belonged to the Vatican Observatory, since they had been supplied for use with the Zeiss double astrograph. The third, with a 65-cm diameter and an angle of refraction of 2.5 degrees, was

purchased from the same firm that built the telescope.

Two reflector guide telescopes, made according to an original design of Junkes, were made in such a way as to permit one to directly observe the field being photographed by the Schmidt when it was equipped with a prism. In 1966, a third guide telescope was added, this one a refractor made in the observatory according to a design by Treanor. It was more suitable than the

Above: The Schmidt astrograph in its dome in the Papal Gardens, with Fr. Martin McCarthy, who used it extensively in the 1960s and 1970s.

Astronauts and Computers Come to the Vatican Observatory

The American astronaut Frank Borman was head of the Apollo VIII mission; in December 1968, together with James Lovell and William Anders, he made the first round-trip to the Moon, circling our satellite 10 times. On his trip to Rome in February 1969, after an audience with the Pope and two formal talks that he gave in the Vatican, he also, at his own request, visited the observatory with his family.

77

Then, on the night of July 20, 1969, came the landing of the first men on the Moon. Accompanied by Fr. O'Connell, the Pope followed this great event from the facilities of the Schmidt telescope as it was transmitted by television. After he had observed the Moon at the telescope, he spoke to the astronauts in these words: "Here, from his observatory at Castel Gandolfo, Pope Paul VI is speaking to you astronauts. Honor, greetings, and blessings to you, conquerors of the Moon, pale lamp of our nights and our dreams! Bring to her, with your living presence, the voice of the spirit, a hymn to God, our Creator and our Father. We are close to you, with our good wishes and with our prayers. Together with the whole Catholic Church, Pope Paul VI greets you."

Meanwhile, astronomical research continued in Castel Gandolfo. The analysis and interpretation of astronomical data almost always require such long and complex calculations that more time is usually required for them than for the actual telescopic

previous ones for guiding the telescope when it was used for long exposures without prisms or for those taken close to the zenith.

The Schmidt telescope was used for about 20 years to study the evolution of star clusters by stellar spectroscopy and by polarimetric measurements.

Above: In February 1969, soon after he captained the first manned spacecraft to orbit the Moon, the American astronaut Frank Borman visited the Vatican Observatory. Here, in the observatory offices connected to the Schmidt telescope, he presented a photograph he took of the Earth as seen from the Moon to the observatory director, Fr. O'Connell.
Top: In 1965, the Vatican Observatory inaugurated its first computer, an IBM 1620, with a visit from Pope Paul VI. Explaining the workings of the computer to His Holiness is the Vatican astronomer Fr. Florent Bertiau.

needed by both the observatory and the astrophysical laboratory. This computer was also used for research on light pollution in Italy.

The remarkable capacity of the new computer caused a reduction in the use of the old IBM 1620, until finally in 1978, thanks to the very rapid progress being made in computer technology, a new computer, an IBM 5100 with a CalComp control unit, replaced both the IBM 1620 and the telephone

link with the Mark I in Milan.

In 1983, the observatory, like all other departments of the Vatican City State, was linked by a terminal to a Honeywell workstation set up in the Computer Center of the Governor's Office. At the same time, the observatory astronomers, who had taken up work in Arizona, began to make use of the computers available at the new location in Tucson. It thus came about that the computers at Castel Gandolfo

observations. And so, towards the beginning of the 1960s, the astronomers of the observatory began to make use of electronic computers. The first to be involved in this new venture was Fr. Florent Bertiau, who wrote programs for the IBM 1620 computer of the University of Rome and the Monte Mario Observatory. In 1963, an IBM 026 card punch machine was purchased in order to facilitate the preparation of programs for the computer. A second machine was acquired a little later, and this was set up at the Schmidt telescope building.

Beginning in 1965, the Vatican Observatory established its own independent Computer Center in the basement of the Schmidt telescope building, and it was inaugurated by Paul VI on August 7. The group of IBM machines included: the central 1620 computer, a 1622 card punch and reader, a 1623 active memory bank, a 1311 disk storage drive, a 047 printing tabulator, two 026 card punchers, a 080 card sorter, and passive memory disks.

In 1971, there was a further improvement in the computer center with the addition of a teleprinter terminal, linked by telephone to the Honeywell Mark I computer in Milan and later by satellite to a still more powerful Mark II in Cleveland, Ohio. A CalComp Plotter 210/563 controlled by this computer provided new possibilities to the Computer Center for the preparation of complex graphics

Above: Patrick Treanor was born in London in 1920, entered the Society of Jesus in 1937, and was ordained in 1952. While at Oxford, he won the Johnson Memorial Prize for his doctoral thesis on interference phenomena. He joined the Vatican Observatory in 1961, and in 1970 succeeded O'Connell as director. His principal areas of research were polarized light from stars and from the interstellar medium (for this, he designed a special rotating polarimeter to be used with the Schmidt); metallicity for various spectral types of stars; faint main sequence stars in the Pleiades; and faint H-alpha emission-line stars. He made important improvements to the Schmidt telescope, including the design and construction of a new guide telescope and, in collaboration with Father Otten, an automatic apparatus for lengthening the images of stellar spectral lines.

were replaced in 1984 by a computer particularly suitable for scientific programs being carried out there. In 1990, the Consortium ICRA (International Center for Relativistic Astrophysics), of which the Vatican Observatory is a member, began to make available at the Castel Gandolfo headquarters a series of ever more powerful workstations.

Today, the needs of the Vatican Observatory are met by the increasingly more powerful computers that are ever more numerable and connected to the Internet Office of the Holy See. The new computers are so powerful that a research project such as the Astrographic Catalog, which required so many years to complete, could be finished in a matter of days.

The Move West

By the 1970s, it was clear that the city lights that had driven the Vatican Observatory out of Rome were now encroaching upon Castel Gandolfo as well. To look for a new site for the observing programs, Treanor, in collaboration with Bertiau and De Graeve, began to research the darkness of the night sky in Italy (for which he designed a special portable photometer). As part of this research, which was also of interest to the Italian Astronomical Society, he served as secretary of the society's commission that dealt with that research at the General Assembly of the IAU at Grenoble in 1976. It was at this assembly that Fr. Patrick Treanor, the observatory's director, presented the official invitation to celebrate the centenary of the death of Fr. Secchi by holding an IAU Colloquium at the Vatican. He organized the colloquium, but unfortunately could not participate because of his unexpected death on February 18, 1978.

Immediately after the death of Treanor, various possibilities were considered for the future of the observa-

tory. The idea of moving the Schmidt telescope to Sardinia was dismissed because, although it offered the best sky conditions in Italy, these did not appear to be such as to justify the trouble and expense of transporting there a telescope like the Schmidt. Other possibilities were, therefore, considered. One was to accept the invitation to transfer the Schmidt to the Canary Islands, where an international observatory was being developed. Another

was to make the Vatican Observatory a member of the European Southern Observatory (ESO), a consortium of European countries with an observing station at La Silla, near the Chilean Andes, in a region where, for varying periods of time, Vatican astronomers had worked in the past. Still another possibility was to accept the invitation to set up a branch of the observatory in Tucson, Arizona, with the prospect of using the large telescopes located there

Above: Fr. George Coyne became the director of the Vatican Observatory following the untimely death of Fr. Treanor in 1978. Born in Baltimore, Maryland, in 1933, he entered the Jesuit order in 1951 and was ordained in 1965. For many years, he did pioneering work at the University of Arizona on the study of dust in space, from planetary surfaces to disks around stars, observed in polarized light. When called to take over the Vatican Observatory, he was director of Arizona's Catalina Observatory and acting chair of the university's astronomy department. This close connection with the University of Arizona facilitated the foundation of the Vatican Observatory Research Group in Arizona in the 1980s.

and to collaborate with many other institutes, which together make Tucson one of the most important astronomical centers in the world.

The appointment of Fr. George Coyne as director of the Vatican Observatory contributed in a decisive way to the choice of this last alternative. He had been a member of that community since 1969 and had for some years conducted research and held teaching positions in Tucson with the several astronomical institutes of the University of Arizona. Over a period of time, he had developed very good relationships with the astronomers at the university — so much so that at the time of his succession to Treanor he was director of the Catalina Observatory and acting director of Steward Observatory, both institutes of the university.

In a similar way, other astronomers at the Vatican Observatory had been involved in research with institutes in the Tucson area. The fact is that for some time, at least among the younger members of the Vatican Observatory community, there was the growing expectation that research would not be carried out exclusively within the confines of the observatory at Castel Gandolfo, but that some permanent collaborative relationship would be established with an important university center capable of offering space and high-quality instrumentation to the Vatican astronomers. And so, the new director,

already himself a member of a very prestigious astronomical institute, considered it best to put off for the moment the project to transfer the Schmidt telescope to Sardinia and to try to set up a collaboration between the Vatican Observatory and Steward Observatory of the University of Arizona in Tucson.

The experiment began in 1980. An agreement was established between the Vatican Observatory and the University of Arizona, whereby, with the payment of an annual fee, the Vatican astronomers in Tucson (who became known as the Vatican Observatory Research Group, or VORG) would be guaranteed both office space and services at the university and also access to the Steward Observatory telescopes. Furthermore, the choice of Tucson provided access to the many large telescopes that other institutes — such as the National Optical Astronomy Observatories, the Smithsonian Institution, and the University of Arizona Observatories — had located in the mountains bordering the desert about Tucson, precisely because of the exceptional quality of the atmosphere and climate for astronomical observations.

The clearest and most tangible sign of the fruitfulness of the collaboration between the Vatican Observatory and the University of Arizona, and in itself the best confirmation of the wisdom of preferring Arizona to other possible places, is without a doubt the Vatican Advanced Technology Telescope (VATT) on Emerald Peak, at an altitude of 3,200 meters, in the Mt. Graham mountain chain, about 160 kilometers northeast of Tucson.

This project, certainly not foreseen at the time the agreement was set up, had come about only because Steward Observatory, chosen collaborators of the Vatican Observatory, already renowned for the high quality of its research and its valuable instrumentation, had in recent years gained a reputation in technological fields, specifically with the invention

of a new method for fabricating large optical mirrors for telescopes.

All of this came about because the University of Arizona, together with the Smithsonian Institution, in the early 1980s had begun to promote the development of an international observatory with advanced technology telescopes. Most of the sites in the Tucson area were completely occupied. So the partners proposed the development of Mt. Graham, and today this has become

Above: *The revolutionary advance that made the Vatican Advanced Technology Telescope mirror possible was a large spinning furnace, built by Dr. Roger Angel of the University of Arizona. Glass melted in this furnace flows into a parabolic shape as it spins. Shown here is the mirror blank, newly removed from the furnace, set up for testing its optical shape. Note the honeycomb voids in the mirror, which reduced the weight of the glass while preserving its strength.*

known at the Mt. Graham International Observatory (MGIO). It is now a reality, but many obstacles had to be overcome in order to see it happen.

It was well known that in order to probe ever more deeply into space to observe astronomical objects ever fainter and more distant, and thereby to look also further back in time, it is necessary to collect ever-greater amounts of radiation from those objects. This is accomplished by increasing as far as

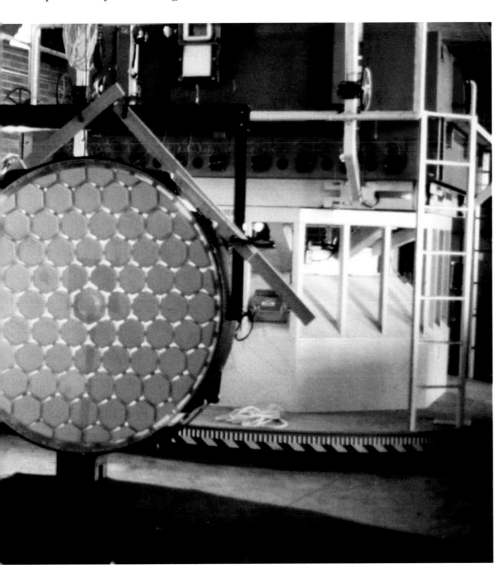

possible the surface area (and, therefore, the diameter) of the telescope mirror, which gathers the light in order to form the image of the observed object.

However, the old technology of making mirrors encountered both technical and financial difficulties in trying to increase the mirror diameter

beyond a certain limit. In fact, the difficulties increase so rapidly with size that they soon become prohibitive. This explains why, after the construction in the 1940s of the five-meter telescope on Mt. Palomar in California and the six meter-Russian one, no other mirrors of such sizes had been made.

Now, however, several new different techniques are competing in the construction of mirrors 10 meters and more in diameter. We have thus entered into the era of the so-called advanced technology telescopes.

As we mentioned above, Steward Observatory is at the forefront in these efforts, having adopted a particular technique for mirror fabrication that uses a rotating furnace. The idea, which originated with Prof. Roger

Angel and his group, reduces enormously the construction time because the glass, melted in a rotating furnace at 1200° C, is distributed by the combined action of gravity and centrifugal force, in such wise as to give to its free surface a concave parabolic shape. A computer controls both the velocity of rotation and the cooling so as to determine very accurately the mirror's shape as it solidifies. The final result is a glass disk that is rigid but relatively light because it is supported from behind by a honeycomb structure. Furthermore, because it is already concave, it does not have to be ground but only polished and surfaced. This allows for remarkable savings in glass and in the polishing time. To appreciate this statement, it is enough to consider that the five-meter Mt. Palomar mirror, begun in 1935, required about 20 tons of pyrex, 25 days for the fusion, 1 year for the cooling, and 12 years in the removal of about 5 tons of glass to obtain its parabolic surface.

The first test mirror, 1.83 meters in diameter, was produced in 1985 as the first experiment in the new technology. It took four hours for the melting process and four for the cooling under rotation. A new technology was also applied to the polishing of the final surface. It is called stressed-lap polishing and can produce an exceptionally exact curvature for short focal length mirrors. Because it was an innovative technique, three years were required to polish the VATT mirror. It was completed in 1991. At the end, a thin layer of aluminum was deposited on the surface.

The end result brought great satisfaction to the University of Arizona Mirror Laboratory, and they were happy to offer to collaborate with the Vatican Observatory in using it for the construction of a telescope to be shared on the basis of the respective contribution of each partner: 75 percent to the Vatican and 25 percent to the university.

The Vatican astronomers could not have received a more welcome gift. It is one thing to have access to telescopes of other institutes with the huge limitation of having to wait in line and then being granted only a part of the observing time requested; it is quite another thing to have one's own telescope with the possibility of carrying out long-range research programs, a privilege not allowed to many research groups.

The Holy See was not able to provide in its budget for the financing of this project, but it encouraged the observatory to accept the offer to build the telescope, if it thought that it would be possible to raise the necessary funds among friends and benefactors.

The promising results of a first campaign in 1987 to gather funds in the United States led to the establishment of the VOF, an autonomous corporation legally recognized in the State of Arizona with the purpose of assuring the necessary means for the construction and operation of the telescope and associated observatory and of funding fellowships for young researchers who wish to collaborate with the Vatican astronomers.

With a second campaign in 1989, a total of $3.5 million was raised, and this was deemed at that time adequate to begin the project. The primary mirror is the property of the University of Arizona. All of the remaining parts of the VATT are the property of the foundation. Since it was the first telescope that employed the technology, and since the major partner was the Vatican, it was given its name: Vatican Advanced Technology Telescope (VATT).

The donors and supporters of this undertaking form the Vatican Observatory Guild. The honor of having their names attached respectively to the telescope and to the adjoining astrophysical facility fell to the lot of Mr. Fred A. Lennon, who asked that the telescope be named for his wife Alice,

and Mr. Thomas J. Bannan, the two major donors. In light of the generosity shown by some American Catholics in wishing to contribute to the new efforts of the Vatican Observatory, we hear again today those words, as it were prophetic, pronounced almost a century ago by Pius X, when, as he crossed the iron bridge at the old Vatican Observatory, he commented: "Ah, these Americans, these Americans!"

On more than one occasion, John

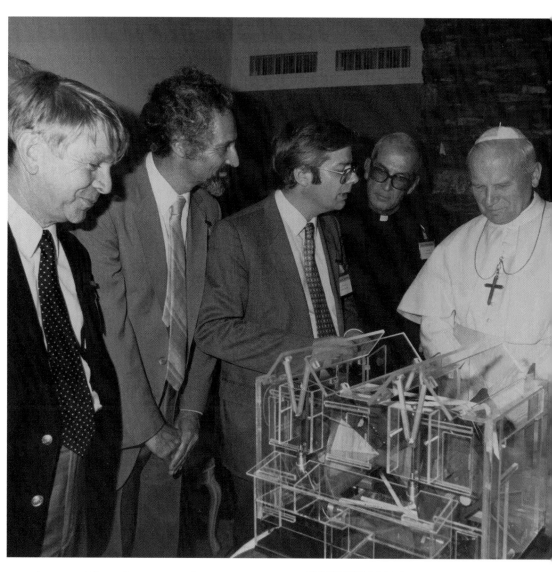

Above: The Vatican actively supported the observatory's efforts to raise the money needed to build the new VATT. During his visit to Arizona in 1987, Pope John Paul II met with astronomers from the University of Arizona planning the new observatory. From left: Peter Strittmatter, Nick Wolfe, Roger Angel, and Fr. Coyne listening in as Dr. Angel describes his plan to the Pope.

Paul II showed his interest in the new undertaking and his gratitude to the American supporters. In September 1987, while on a pastoral visit to Phoenix, Arizona, he did not forget his astronomers residing in nearby Tucson. In fact, he received a delegation of scientists from the Vatican and from the University of Arizona who, led by Coyne, described the new telescope project to him. In June 1989, he also received some of the members of the Vatican Observatory Guild in a special audience.

The Vatican Advanced Technology Telescope

The VATT, which was inaugurated with celebrations on September 17-19, 1993, has essentially two parts: the Alice P. Lennon Telescope and the Thomas J. Bannan Astrophysics Facility. The building contains the control center, the computer station, offices, and a four-bedroom apartment to accommodate the astronomers. The telescope is isolated both mechanically and thermally from the rest of the facility in order to avoid deterioration of the telescope images by vibrations or by heating, even that caused by humans.

As to the optics, the primary mirror has a focal length equal to its 1.83-meter diameter and, therefore, a focal ratio that, at least for mirrors of this large size, is altogether exceptional,

and, in fact, unique. This fact makes it very compact and somewhat like a cube, which, in addition to giving the telescope an unusual look, provides a large financial savings both in the telescope construction and in the building and dome that house it, since they, too, can be smaller than normal.

The aluminum Ash dome is 7 meters in diameter and has a shutter opening of 2.5 meters, an unusual relative size for the opening. The reason for this is to reduce to a minimum perturbations by the wind and by thermal differences within the dome and between the dome and the outside ambient. This contributes a great deal to the image quality.

The primary mirror of the new telescope is made of borosilicate (Pyrex) glass, backed by a honey-combed structure which makes it both rigid and light. This structure, at the same time, provides for a rapid and uniform response to changes in the ambient temperature — and, in fact, air is circulated within the structure to help avoid deformations due to temperature differences.

The images formed by the telescope are not observed visually, nor are

Above: The final stage in the preparation of the mirror is the final polishing of its surface. Inspecting the mirror is Buddy Martin of the University of Arizona. With careful testing, the VATT telescope mirror was able to be polished to an accuracy unprecedented at that time.
Top: Once the molten glass has been cooled, it is fixed into the ideal shape for a telescope mirror. Note the honeycomb structure visible here: the molten glass was formed into the mirror shape over a series of ceramic blocks. Once the blocks are removed, one obtains a mirror disk that is both strong and lightweight.

they photographed. Rather, they are detected electronically using a charge coupled device (CCD) chip, displayed on a computer monitor, and recorded on disk.

The telescope is the Gregorian type in that it uses a concave secondary mirror made of zerodur glass. This too is an uncommon characteristic, which provides advantages both from the optical and the financial point of view. The secondary provides an effective focal ratio of F/9 and is located beyond the primary focal point. With this focal ratio, we have a good relative aperture without a long telescope. Thus the field of view is larger than one would have with a classical Cassegrain focus. In fact, it is 15 arc minutes, about one half the size of the Moon for a curved field and 10 arc minutes for the flat field seen by the CCD.

As to its resolving power, its ability to obtain well-detailed images and to detect faint objects, thanks to its advanced technology and to the site, it today provides, in favorable atmospheric conditions, an image resolution of 0.7 arc seconds, two times better than that of a previous generation telescope of equal size. This also means that it is four times more efficient in detecting faint objects. An even better resolution of about 0.5 arc seconds will be obtained when adaptive optics are employed.

The image quality of the VATT is due also to its mechanical rigidity against the wind and thermal distur-

bances. It is compact because of the f/1 primary and the altazimuth mounting that is similar to a ship's cannon. The direct linkage of large motors to the vertical and horizontal motions also contributes to the stability. These motors are regulated by a computer so that they can point automatically and follow celestial objects with high precision.

All of the technical successes of the VATT have made it a prototype for a whole series of the next generation of much larger telescopes. In fact, the rotating oven and stressed-lap have now been applied to the production of larger mirrors: 3.5 meters, 6 meters, and the two 8.4-meter mirrors for the Large Binocular Telescope (LBT), recently constructed on Mt. Graham near the VATT.

The Science-Faith Dialogue

Writing in L'Osservatore Romano *on February 17, 1965, future director Fr. Patrick Treanor, S.J., commented on how the Vatican Observatory serves as an instrument of dialogue between scientists and philosophers. His ideas form the basis of the following reflection:*

At the inauguration of the new headquarters at Castel Gandolfo, Pius XI had done no more than reemphasize the ideas enunciated 50 years before by Leo XIII. He said that the purpose of the Vatican Observatory was "to assure for the Faith and for Religion that implicit, rather explicit, support which shines forth and is more than ever persuasive each time that respect for the Faith is shown to be united in a fraternal embrace with the cultivation of science."

The goal, therefore, was still to establish and facilitate the dialogue between science and faith, a goal that with each passing year has become ever more timely and vital in the Church's self-awareness. In fact, from those times to today, science as a whole and in particular astronomy have made spectacular progress, and the importance of science in human society is ever increasing.

At the same time, the climate of renewal that has animated the Church since the Second Vatican Council and

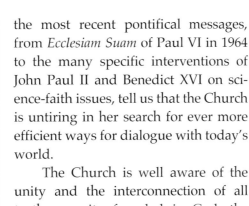

the most recent pontifical messages, from *Ecclesiam Suam* of Paul VI in 1964 to the many specific interventions of John Paul II and Benedict XVI on science-faith issues, tell us that the Church is untiring in her search for ever more efficient ways for dialogue with today's world.

The Church is well aware of the unity and the interconnection of all truth, a unity founded in God, the author of all truth. Under the influence of materialism and skepticism, modern man has, now more than ever, lost for the most part this sense of unity. Faced with the rapid growth and the obvious power of science, he tends to find in the scientific method an alternative to the Christian philosophy of life. Even where there is no intentional opposition or conscious prejudice, the barriers of a different cultural substratum and also of a peculiar technical language isolate an ever-growing part of mankind from the message of the Church,

with the danger that the Church herself remains in turn isolated from what science has to say.

[In this regard, it is important to note that John Paul II, in repeating how necessary the science-faith dialogue is, also put the emphasis, more so than any of his predecessors had done, on the fact that the dialogue cannot exist as a one-way street. If it is true that science must be open when theology seeks to approach it, it is no less true that theology must be open and attentive to progress in science.]

It must be said, to be truthful, that modern cosmology — that part of astronomy which has undergone huge developments in the past decades as it explores the physical origins and evolution of the universe — offers today an altogether new and special opportunity for the exchanges that the science-faith dialogue imply.

Cosmology, in fact, more than any other branch of science, brings the scientist to deal with problems and formulate conclusions that, as they touch upon interdisciplinary matters, lead in a natural way to ultimate questions that are no longer only about physics but rather about metaphysics, and not infrequently about theology. And so today, scientists on the one hand and philosophers and theologians on the other find themselves being challenged anew to mutual exchanges where they each set out to meet the other, knowing that they must together both give and receive.

In this perspective, the way in which the Vatican Observatory sees its mission in today's world becomes clear. The first object is to become a part of the intellectual life and the ongoing development of modern astronomy in a clear productive manner. It is not a matter of creating an image that gives the appearance that the Church is interested in science, but rather to do science in the fullest sense of the word. Through this work one hopes to

Secondary Mirror
D = 0.377m
f/ = 0.8773
k = −0.655
Sag = 0.0274m
Asp = 603waves

2.201m

Primary Mirror
D = 1.83m
f/ = 1.00
k = −0.9958
Sag = 0.1144m
Asp = 2875waves

1.143m

Focal Surface
Field Diam. = 15 arcmin
Field Scale = 12.52 arcsec/mm
Rad of Curv = −0.34m
System f/ = 9.0

Above: The VATT in its dome on Mt. Graham, Arizona.
Right: Technical specifications of the VATT optics. The unusual optical path, where the light reflected from the primary mirror reaches a focus before encountering the concave secondary mirror, is called a Gregorian design. It allows for a more precisely shaped secondary mirror, thus improving the overall accuracy of the telescope; but keeping the telescope in focus requires extremely precise positioning of the secondary, done by computer. The VATT was not only the world's first large telescope with a spin-cast mirror, but also the first of its size with Gregorian optics. Choosing such a design was a kind of "Gregorian chance" which has paid off beautifully!

instill in the modern world, Catholic and otherwise, an exact evaluation of the place of science in Christian life and thought and to help colleagues in science to recognize the value, no longer only physical but philosophical and theological, of the ultimate questions to which they, more and more naturally, are led by their research in physics and, in particular, in cosmology.

Competition or Collaboration?

But how can Vatican astronomers pretend today to be up to doing serious and competitive scientific work when there is an always increasing number of institutes that rely for their research on space telescopes or on the new observatories being constructed around the world that are equipped with instruments enormously more powerful and costly than those of the Vatican Observatory?

In this regard, we should remember that modern astronomy covers a vast variety of research programs, and that

FR. SABINO MAFFEO, S.J. (Italy), is the archivist of the Vatican Observatory. This chapter was adapted from his book, Specola Vaticana: Nove Papi, Una Missione (The Vatican Observatory in the Service of Nine Popes), translated by George Coyne.

each of them has need of instruments with special characteristics, of which size is but one, and not always a necessary or sufficient one. The most powerful telescopes must carry out research at the very limits of their great potential, and they can only do this well if they have at their side smaller telescopes fitted for research that is equally fundamental and fascinating and for which the large telescopes would be poorly employed. The relationship, therefore, between large and small observatories is such that they complement rather than compete with one another.

It is worth the trouble, in this regard, to remember how totally valid today are the words that Treanor wrote in 1973 with respect to the challenge facing the Vatican Observatory due to the enormous development of astronomy in recent decades:

It would be an illusion for us to put ourselves on an equal footing with the large institutes which have almost unlimited technical and economic resources. The world offers us many other examples of observatories, relatively small but good, whose contribution to scientific progress exceeds by a large measure their physical dimensions.

We have inherited from Fr. Secchi a tradition that has carried the Vatican Observatory's interests toward ever-deeper studies of our stellar system — that is, the Galaxy. It is a field in which our astronomers are already known for their expertise, and for which our instrumentation (thanks to a strategic updating initiated during the last decades) is particularly suitable. Above all, it is a field of astronomy in rapid growth where observatories, even those with limited resources, can make a notable contribution to the observational techniques and interpretation on which all progress in astronomy is ultimately based.

With the installation of the advanced technology telescope on Mt. Graham in Arizona, the Vatican Observatory has brought its instrumentation up to today's standards,

thus taking a place among those observatories equipped with instruments of modest power, the larger number by far in the world.

A New Way of Doing Research

The fact that there are now two locations, Castel Gandolfo and Tucson, marks a new stage in the history of the Vatican Observatory, a new way of doing research. Characteristic of this new way is international collaboration, ever more widespread and at various levels — so much so that the work of the Pope's astronomers is today, more so than in the past, carried out on a worldwide basis.

In fact, the Tucson venture, on the one hand, has nurtured the real scientific research development of the Vatican astronomers and their collaboration with the prestigious institutes located in Tucson. On the other hand, quite contrary to what one might have thought, there has been at Castel Gandolfo the development of new, intense, and flourishing initiatives on an international scale. These include meetings and summer schools (see pp. 116-119), which had never been held there before.

One might say that the new program represents a way whereby the staff of today's Vatican Observatory carry out their work in tune with the vision of the modern Popes to promote dialogue with all peoples.

From the beginning of the 1980s, with the arrival of some new Jesuit astronomers at the observatory and with the opening of the VORG in Tucson, Arizona, there began, in addition to continued studies on our own Galaxy, research on external galaxies and theoretical studies in planetary sciences, astrophysics, and cosmology.

A description of some of that work can be found in the following chapters of this book. ●

STARS AND GALAXIES

Above: The Milky Way appears as a streak of light across the sky in this image taken from Mt. Graham in the 1980s. Note the dust bands obscuring the central plane of the galaxy. The picture was taken to test the skies where the VATT and Large Binocular Telescopes were eventually built. The yellow glows of light on the right are the city lights of Tucson, 75 miles away, and Phoenix, more than 100 miles away.

STARS AND THE MILKY WAY

• Fr. Christopher CORBALLY, s.j. •

The Census of the Stars

When we are well away from a town or city and look up at the night sky, we can be overwhelmed by the myriads of stars over us. Through a telescope, the sight can be even more inspiring. But how can we find out more about the personalities of these stars?

Everything we know about stars we have learned from their light, which gratuitously falls on the Earth. It was Newton who discovered that light can be split up by a prism. He did this for our nearest star, the Sun, and found he was looking at the colors of the rainbow.

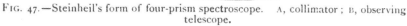

Fig. 47.—Steinheil's form of four-prism spectroscope.   A, collimator ; B, observing telescope.

Left and above: *Diagrams of 19th-century spectroscopes, taken from J. N. Lockyear,* Contributions to Solar Physics, *published in London by MacMilland and Co. in 1894. This book was a gift to the Vatican Observatory from William Lockyear, inscribed February 1932.*

Now these colors are beautiful, but the really interesting details contained within a rainbow, or "spectrum," are revealed when light from a star is focused onto a narrow slit, which from there passes through a prism, and then gets focused again onto your eye or a camera.

A History of Stellar Spectra

Josef von Fraunhofer first split up the light from a star in 1814. At first, he had a tiny telescope, so he could only look at the brightest star in the night sky, Sirius. Later he used a larger telescope and observed more stars. But of course the Sun, as the very brightest "star," gave him the best spectrum in which he was able to see the dark "lines" where light was

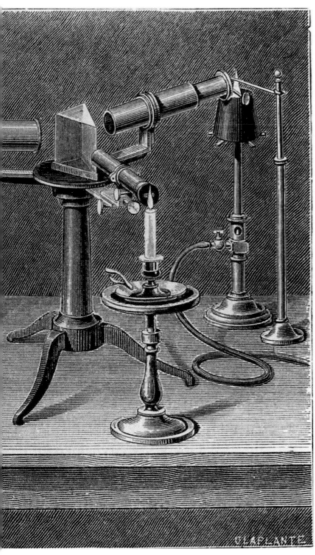

missing from the rainbow.

Later it was understood why some of the light was missing. While all of the colors of the rainbow are made in the interior of the Sun, its atmosphere, being cooler than the interior, absorbs the out-streaming light at very specific colors, or wavelengths, that correspond to the atoms and molecules there. Each atom or molecule has, as it were, its own unique "fingerprint" in which the strengths of the features corresponds with the particular conditions of temperature and pressure of the gas making that fingerprint or pattern of lines. So the spectrum of a star tells us what is in its atmosphere, and how hot and big the star is.

It took nearly 50 years after Fraunhofer observed the first stellar spectrum for the next big breakthrough. In 1863, Giovanni Battista Donati (Italy), George Airy, William Huggins (both in England), Lewis M. Rutherfurd (United States), and Fr. Angelo Secchi (Vatican) each published pioneering papers on their observations of stars through a spectrograph. Our knowledge of stars really took off from that time!

Fr. Secchi, after travels in the United States, returned to Italy and became director of the Roman College Observatory, the predecessor to the current Vatican Observatory. Besides spectroscopy, his wide interests covered meteorology, terrestrial magne-

tism, sunspots and other solar chromospheric phenomena, double stars, and comets. By the end of his life, he had sorted over 4,000 stars according to their spectra. He arranged them into four classes, and he thought about the differences in physical conditions that would be bringing about the differences he was seeing in their spectra. While Huggins might be regarded as the founder of stellar spectroscopy, Secchi's pure and prolific approach makes him the father of stellar spectral classification, along with the branches of astrophysics that his methods encouraged.

The sorting of stars according to their spectra continued to be refined through such as Hermann Carl Vogel's work in Germany, first in Bothkamp and then Potsdam. From 1885 and for the next four decades, a major center of spectral classification was the Harvard College Observatory in Cambridge, Massachusetts. There, the energetic director, Edward C. Pickering, encouraged the sharp eyes and expertise of Williamina Fleming, Antonia Maury, and Annie Jump Cannon, with their

Above: Edward Pickering stands in a room filled with his "computers," who are classifying stars and producing the Harvard College catalogs. Also standing, in the middle, is Williamina Fleming, while Annie Jump Cannon, on the right, peers at spectra through a microscope. The photo was taken in 1892, and is courtesy of the Harvard College archive.

fellow "computers" (as these assistants were called), to produce huge catalogs of stars. So dedicated was Cannon that, by the end of her life, she had classified over 395,000 stars in the "Draper system" of the Harvard College Observatory, which not surprisingly had become the internationally accepted system at the first General Assembly of the International Astronomical Union, in Rome, in 1922.

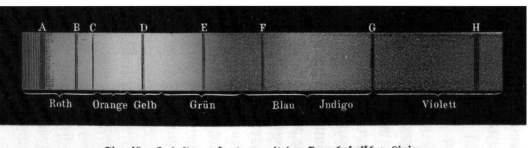

Fig. 49. Das Sonnenspectrum mit den Fraunhofer'schen Linien.

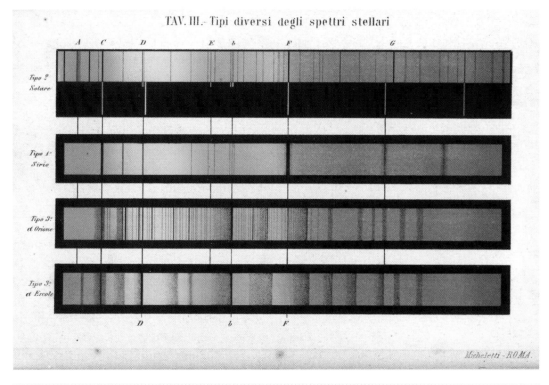

TAV. III.- Tipi diversi degli spettri stellari

Top: The Sun's spectrum with the Fraunhofer Lines. These dark features show where light is missing from the full rainbow of colors due its absorption by atoms and molecules in the outer atmosphere of the Sun.
Bottom: Fr. Angelo Secchi's drawings of "Different Types of Stellar Spectra" are from his by-eye observations of four bright stars. The top one, of the Sun, illustrates his spectral class 2, and it is marked with features first labeled by Fraunhofer. The next, class 1, is illustrated by Sirius, the hottest star in this picture. The two spectra at the bottom are both of class 3, but α Herculis (Rasalgethi) is a bit cooler than α Orionis (Betelgeuse) and so has banded features that are more pronounced. These bands are due to titanium oxide molecules. Secchi's fourth class is not illustrated here, but he deduced that its bands of "reversed intensity" were due to carbon molecules. Rasalgethi and Betelgeuse are huge, supergiant stars, while the Sun and Sirius are regular-sized stars, called dwarfs in comparison.

Spectra and Brightness

In the 1920s, Walter Adams and Arnold Kohlschütter, while working at the Mt. Wilson Observatory in California, noticed that besides changes with temperature, some spectral features changed depending on the overall size and so brightness of the star while other features did not. The ratio of the changing to the unchanging features, or "lines," allowed a calibration of such ratios with the star's actual brightness. And knowing how bright a star actually is, its "absolute brightness," and also how bright it appears to us, gives a measure of the star's distance away from us. (Just as knowing at night whether it is a torch or a searchlight shining at you gives you an idea of how far away it is.) Hence was born the technique of "spectroscopic parallaxes," or distances to stars. With that, we can plot out how different stars in our Milky Way Galaxy are distributed. When it was also realized that the very hottest stars were the youngest, one could pinpoint the regions in our Galaxy where stars have most recently been formed. "Galactic structure" work was now possible, leading to an understanding of what kind of stars make up our Milky Way, particularly in its spiral arms, and so outline our Galaxy's history.

Obviously, the better one could pin down the luminosity of a star, the better that galactic structure work would become. The team of William W. Morgan and Philip C. Keenan, while together at Yerkes Observatory in Wisconsin, so carefully refined the way in which stars of different surface

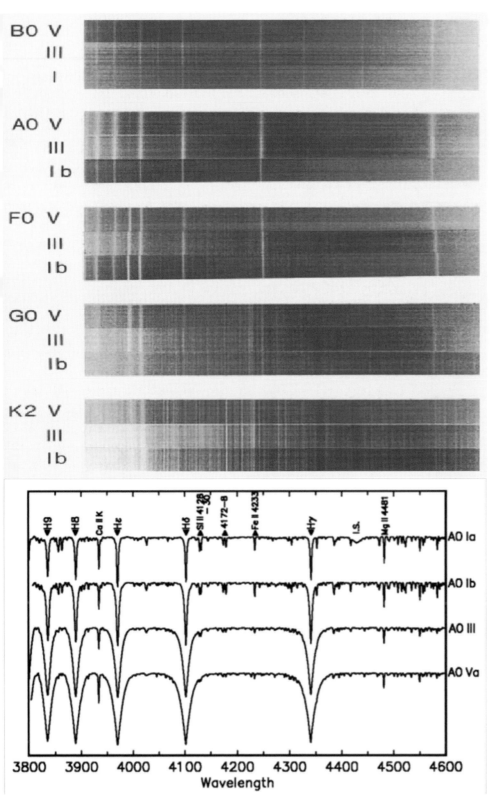

Above: Sets of spectra of stars with different brightness, ranging from the dwarfs "V," through the giants "III," to the supergiants "Ia or Ib." Above are the classic photographic spectra, shown as negatives. Notice how from the set of hottest "B" stars at the top to the coolest "K" stars at the bottom, it is the change in temperature that makes the difference in appearance, for the amount of the elements in their atmospheres (hydrogen, helium, carbon, iron, etc.) does not change. Below are spectra observed with a modern digital detector. Their downwards-directed "teeth" correspond to the light lines in the photographic spectra and are for a set of "A0" temperature stars, like the second set in the photographic spectra above. Many people find the details in the digital spectra clearer than in the photographic ones, and certainly one can now reach fainter stars with digital detectors.

temperature could be classified into their respective luminosities that this "MK system," with its unchanging lists of standard stars, is still the way in which it is done today. In producing their beautiful atlas of spectra, showing the effects of luminosity, they were helped by Edith Kellman; their definitive work published in 1943 is known as the MKK Atlas.

It happened that a full 13 of the MKK Atlas's printed plates of spectra showed stars that had peculiarities in relation to the normal run of stars. So it was clear even then that further refinements were needed to the MK System. This identification of peculiarities and the attempt to discover the physical reasons for them has become the bread-and-butter work of many stellar spectral classifiers since then.

Classifying Stars

Classifying stars is a bit like classifying people. At sufficient distance, everyone looks the same. Closer, and one sees the major differences between them, and one can sort them into classes depending on height, apparent age, color of skin, and prominent features. Very close up, and everyone is different. The trick for classifying star spectra is to get sufficient detail for a general description of a spectrum while

Above: Hubble Space Telescope image of the star cluster NGC290. The bluest stars are the hottest; the bright reddish stars have evolved beyond their "middle age" period into red giant stars. Yellowish stars like our Sun are among the fainter ones. Credit: *European Space Agency & NASA*. Acknowledgment: *E. Oiszewski (University of Arizona)*

not being overwhelmed by the differences between the spectra. Morgan and Keenan got this "right distance," or spectral resolution, perfect for stars.

They also used spectra that showed the blue-green region of the spectral rainbow. This was because their photographic emulsions were the most sensitive in this spectral region, but it also turns out that the "fingerprints" of atoms and molecules vary the most strongly there with small differences in the temperatures and pressures of the stellar atmospheres. Modern work in stellar classification has taken advantage of the different spectral regions

that new detectors and observing possibilities from satellites have brought. So the ultraviolet, available only in space, is best for studying stars with huge winds blowing from their surfaces or for identifying the hot companion of a twin star system; the infrared will work best for objects that are very cool, like "brown dwarfs," or for stars that are embedded in giant clouds of dust that block the bluer light from getting out to us.

Above: *Blue stars tend to be young stars, and so they are most likely to be found in the spiral arms of a galaxy. The distribution of young star clusters and individual blue stars within our own Milky Way Galaxy, as deduced from their known brightness and so distances, can be used to map out a picture of our Galaxy, shown here as if viewed from the "top."* **Credit:** *NASA/JPL-Caltech/R. Hurt (SSC)*

Getting to Know Our Neighbors

We have seen how the spectral classification of stars helps us map out the distant structure of our Milky Way Galaxy. It also helps us get to know our neighbors. And you never know, we might find some that are intelligent! This was the thinking behind the Nearby Stars census of NASA, in which the Vatican Observatory participated. Through spectral classification of all the stars near us, and then comparing their spectra with computer-generated models of star spectra, we can draw up a list of our neighborhood stars that are most like our Sun. Since we know that the conditions that formed our Sun were also suitable to form our planets, the expectation is that these solar-like stars, which we picked out, could also have their own planets around them.

"Most like" also includes the idea that a star has an activity level like our Sun. Young children have bundles of energy, while an adult generally moves more sedately; similarly, a mature star like our own has slowed down from its initial high speed of rotation, and so its surface activity has quieted, though solar flares and prominences can still be impressive on our current Sun. High solar activity in the past generated energetic particles that hit the Earth and made things more difficult for life to form, so the thinking goes. If we are looking for a planet with life around another star, then the less active that star, the better the chance we have of success.

The Simple Picture Gives Way to Surprises

We often start with simple pictures of how things work, and then begin to find the complications. The same is true for our picture of the Milky Way. We have mentioned how its spiral structure shows where stars have recently been formed. In the central, nuclear bulge of the Milky Way are the older stars, including a monster black-hole that has been living happily on a diet of stars that it manages regularly to suck into itself. In the outer halo of the Milky Way, extending from both sides of the relatively flat spiral struc-

Above: The limb of our Sun in extreme ultraviolet light emitted by ionized helium, an element which was first discovered in the Sun. The flaring of the prominence and the glow from its outer atmosphere show its amazing activity. Yet compared with younger stars, this activity is quite low. The recent Census of Nearby Stars measured the activity level of our neighboring stars. Those with low activity are presumed most likely to favor life on any Earth-like planets they would have. Credit: STEREO Project, NASA

ture of the disk, were found the remnants of the initially spherical Galaxy before it collapsed down into a slowly spinning disk. These remnants included globular clusters of stars as well as individual, free-moving stars.

As astronomers were able to find and probe the character of fainter and fainter stars out in the Milky Way's halo, it became clear that collections of these stars formed streams, all moving together through the halo. Where were they coming from? It was realized that these streams of stars are remnants of our own Milky Way's collisions with other, smaller galaxies. This more complicated picture, which we shall explore in the following chapters, of galaxies interacting with each other rather than staying isolated universes, has its traces in the faintest structures of our own Milky Way. When we get to know stars through their spectra, there are always surprises waiting for us. ●

Above: The Star Streams of NGC 5907. This "Knife Edge" galaxy is surrounded by ghostly streams of stars. They have come from a smaller, "dwarf galaxy" whose stars were gradually torn apart from each other in its encounter with NGC 5907, some four billion years ago. Image courtesy of R. Jay Gabany (Blackbird Observatory), D. Martínez-Delgado (IAC, MPIA), J. Peñarrubia (U. Victoria), I. Trujillo (IAC), S. Majewski (U. Virginia), M. Pohlen (Cardiff).

Fr. Chris Corbally, S.J. (Britain), is an expert on the classification of stars, and the co-author of a recent book on stellar spectral classification. This chapter, based on that work, was written specially for this book.

A GALACTIC UNIVERSE

• Fr. JOSÉ GABRIEL FUNES, s.j. •

*"I see myself as a child
playing on the seaside
and sometimes having fun
discovering a small rock or a shell
that is more beautiful than normal,
while in front of me,
the huge ocean of truth
extends unexplored."*

Isaac Newton

358   Mr. HERSCHEL on the

straggling stars of course will be very few in number; and therefore the ground of the heavens will assume that purity which I have always observed to take place in those regions.

Enumeration of very compound Nebulæ or Milky-Ways.

As we are used to call the appearance of the heavens, where it is surrounded with a bright zone, the Milky-Way, it may not be amiss to point out some other very remarkable Nebulæ which cannot well be less, but are probably much larger than our own system; and, being also extended, the inhabitants of the planets that attend the stars which compose them must likewise perceive the same phænomena. For which reason they may also be called milky-ways by way of distinction.

My opinion of their size is grounded on the following observations. There are many round nebulæ, of the first form, of about five or six minutes in diameter, the stars of which I can see very distinctly; and on comparing them with the visual ray calculated from some of my long gages, I suppose, by the appearance of the small stars in those gages, that the centers of these round nebulæ may be 600 times the distance of Sirius from us.

In estimating the distance of such clusters I consulted rather the comparatively apparent size of the stars than their mutual distance; for the condensation in these clusters being probably much greater than in our own system, if we were to overlook this circumstance and calculate by their apparent comprehension, where, in about six minutes diameter, there are perhaps ten or more stars in the line of measures, we should

Above: In 1785, writing in the Transactions of the Philosophical Society *of London, Herschel described his "construction of the heavens." Based on his estimation of the number and distance of the stars around our Sun, he deduced that we inhabited a finite disk-shaped cloud of stars. And he proposed that other nebulae visible in his telescope were their own "milky-ways."*

A Galactic History: Island Universes

It was not until the 20th century that we confirmed that we live in a galaxy, or "island universe," according to Immanuel Kant's hypothesis. The German philosopher was the first in 1775 to propose that spiral nebulae were analogies of our own Galaxy, the Milky Way. Ten years later, Sir Edmund Herschel backed up this idea with his estimate of the size and shape of the Milky Way, suggesting that "other very remarkable Nebuale … are probably much larger than our own system; and, being also extended, the inhabitants of the planets that attend the stars which compose them must likewise perceive the same phenomenon. For which reason they may also be called milky-ways…." Indeed, our modern term *galaxy* is based on the Greek word for milk.

In the early 20th century, however, the nature of the spiral nebulae (galax-

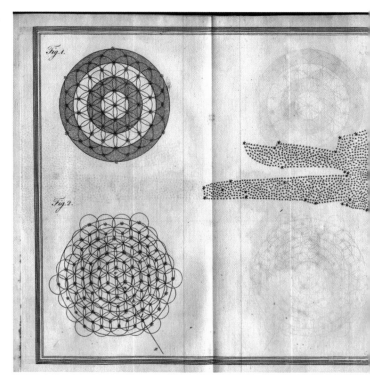

ies) and the size of the universe were the subject of considerable controversy. At the time, nobody was certain whether the spiral nebulae were merely clouds of gas within our own galaxy. "The Great Debate" between Harlow Shapley and Herber D. Curtis was held in 1920 at the National Academy of Sciences in Washington, D.C. Shapley proposed a model for our Galaxy in which the position of the Sun was exocentric, overestimating the size of the Milky Way, practically identifying the Galaxy with the universe. In this model, the spiral nebulae were not "island-universes" but nebulae within the Milky Way.

On the opposite side, Curtis believed the Sun was at the center of the Milky Way and maintained that spiral nebulae were systems with similar dimensions to our Galaxy. Both astronomers knew that the key to solve the dispute was the measurement of the distance to the spiral nebulae.

In 1924, Edwin Hubble proved that the Andromeda nebula was a galaxy similar to ours. Using Cepheid stars, a type of pulsating star, as standard candles, Hubble was able to measure the distance to Andromeda. The Milky Way and Andromeda have a diameter of about 100,000 light-years, and the distance between the two galaxies is about 2.5 million light-years.

In those years, Bertil Linblad and Jan H. Oort discovered that the stars of the Milky Way rotate around a center that was not the Sun's position. The dis-

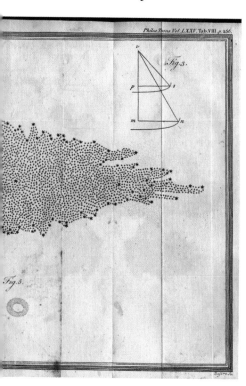

tance from the Sun to the galactic center is about 25,000 light-years. They also found out that the stars in the Milky Way are distributed in a disk and a bulge, both components immersed in a spheroidal halo. Thus it was confirmed that the Milky Way had a structure similar to that of other external galaxies.

What Is a Galaxy Made Of?

Extragalactic astronomers (those astronomers who study galaxies external to our Galaxy, the Milky Way) may feel just like children playing in the seaside of a huge ocean. If we think of the universe made of a hundred billion galaxies, it is easy to realize that we are facing a huge ocean. This number is mind-boggling. To render the number more comprehensible, let's think of the population of the world, which is about 6.5 billion. If we divide the number of galaxies in the universe by the number of people in the world, each of us could have 15 galaxies, each galaxy containing from millions to hundreds of billions of stars.

But are galaxies made only of stars? No, they aren't. Then, what is a galaxy, and what is it made of?

A galaxy is a system of stars, gas, dust, and dark matter. All of this material is gravitationally bound together with a mass ranging from 10 million to

1,000 billion times that of the Sun. The stellar component is distributed in a spheroidal component (the bulge and the halo) and in a flat component (the disk). Gas and dust are the material between stars, and it is called "interstellar medium"; this is the material from which new stars form. As for "dark matter," we don't know yet its nature; but we have detected and weighed the dark matter, and we know as well that it does not emit light. We do know that dark matter is located in the galaxy halos.

These galaxy components (stars, interstellar medium, and dark matter) vary from galaxy to galaxy and define the morphology — the shape — of a galaxy. Galaxies have a disk component and an spheroidal component. Stars in the disk are bluer and "younger" than stars in the spheroidal components.

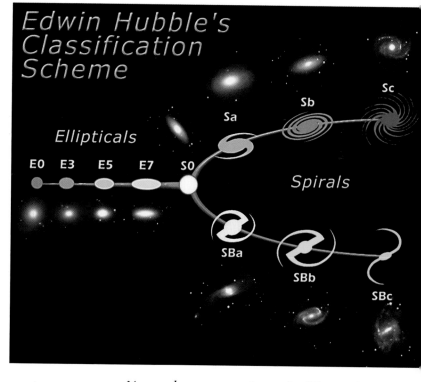

Young, here, means tens of millions of years!

The famous astronomer Edwin Hubble, in 1926, classified galaxies in spirals and ellipticals. His diagram is known as the "Hubble tuning fork." Spiral galaxies are divided in two groups: normal spirals and barred spi-

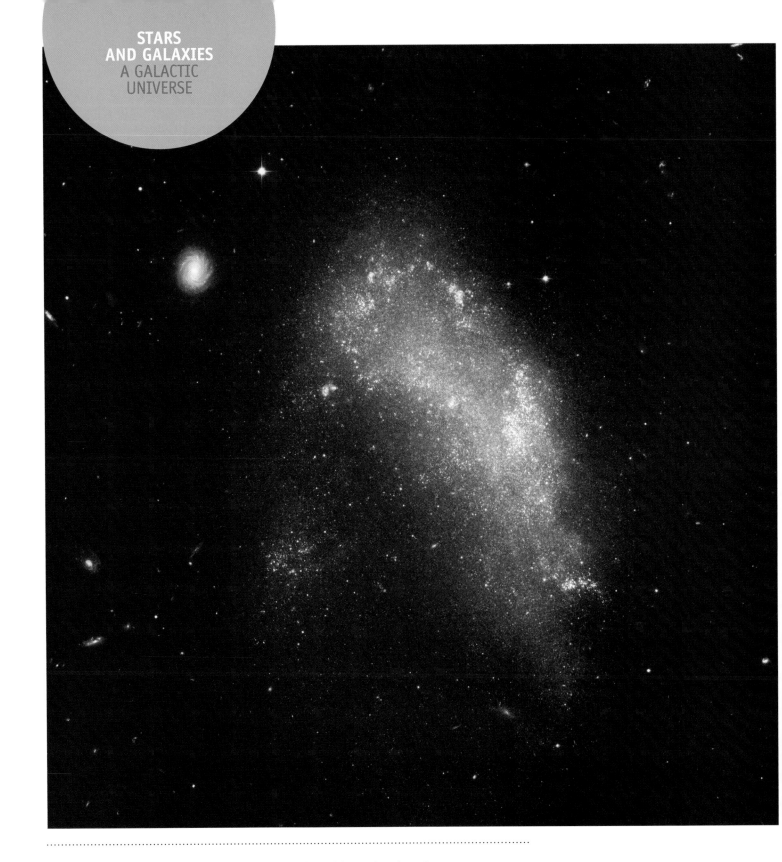

Above: Galaxy morphologies illustrated. In the foreground is NGC 1427A, an irregular galaxy in Fornax. To its upper left is a background galaxy (which happens to lie in the telescope's line of sight but is some 25 times further away), which is a spiral galaxy somewhat similar to our own Milky Way. At even greater distances, background galaxies of various shapes and colors are scattered across this Hubble image.
Credit: *NASA, ESA, and The Hubble Heritage Team (STScI/AURA).*
Acknowledgment: *M. Gregg (Univ. Calif.-Davis and Inst. for Geophysics and Planetary Physics, Lawrence Livermore Natl. Lab.)*

rals. Spiral (disk) galaxies are those in which the disk component is predominant. Using a familiar image, they look like a fried egg. Our own Galaxy, the Milky Way, is a disk galaxy. Elliptical galaxies look like a rugby ball.

Hubble also noticed that there is a small fraction of galaxies that can be grouped in a third major type he called "irregular galaxies." Galaxies in this class don't seem to have any rotational symmetry as Hubble pointed out (see NGC 1427A).

In the preface of his famous book *The Realm of the Nebulae,* published in 1936, Edwin Hubble wrote that "the book is believed to furnish an authentic picture of a typical case of scientific research in the process of development." That statement is still true in 2009, and the Vatican Observatory has contributed and continues to contribute to the progress in this field. Two international conferences on the formation and evolution of disk galaxies were organized by the Vatican Observatory in 2000 and in 2007.

Galaxy Formation

One of the major questions in astrophysics regards the formation and evolution of galaxies. As was pointed out by Sandy Faber (University of California) in the conference summary of the first of these Vatican meetings, galaxies are the crossroads of astronomy because they look up to cosmology and they look down to the interstellar medium and star formation. The study of galaxies is crucial when trying to connect our knowledge of the universe as a whole with the formation of stars and planets.

And so, concerning the question of how galaxies form and evolve, we begin by asking: What causes galaxies to look different?

Our best models for galaxy forma-tion assume that in the first million years of the universe, matter originally filled all of space "almost" uniformly. The distribution of matter was not completely uniform. There were regions that were slightly denser than the surroundings. In these "little" cosmological perturbations, gravity pulled in surrounding matter. Cold dark matter caused matter to fall towards dense regions and made rarefied regions more rarefied. Denser regions contracted, forming protogalactic clouds. Hydrogen and helium gases in these clouds formed the first stars. The first primordial stars began within a billion years after the Big Bang as tiny seeds (with a mass 1 percent of that of the Sun) that grew rapidly into stars 100 times the mass of our own Sun. The supernova explosions from the first stars heated the surrounding gas slowing the collapse of the protogalactic cloud and the star-formation rate. The leftover gas settled into a rotating disk. This simple picture explains the existence of a bulge and a disk in spiral galaxies. The spheroidal component was formed during the collapse of the protogalactic cloud, while the disk component was formed when the galaxy's rotation became organized.

Models of galaxy formation suggest that galaxies look different because the conditions of the galactic clouds were slightly different or because they are the result of the interactions with other galaxies. In other words, galaxies are also the result of a trans-formation.

In the first case, the initial conditions of the protogalactic cloud that could explain the existence of elliptical and spiral galaxies are spin and density. If the protogalactic cloud had a considerable rotation, the cloud would end up in a spiral galaxy. If the cloud rotated very slowly, the final result would be an elliptical galaxy.

Density is the other physical property that rules the evolution of the protogalactic cloud. Higher density causes a faster cooling, allowing more rapid star formation. If the star formation were fast enough during the collapse, most of the gas would form stars, leaving very little gas to settle in a rotating disk. The final result would be an elliptical galaxy. If the cloud has a lower density, it would end up as a spiral galaxy.

Things are a bit more complicated, however. Galaxies are not exactly island universes, in the sense that they don't evolve in isolation; they interact with other galaxies.

Spiral galaxies tend to collect into groups of galaxies, which contain up to several dozen galaxies. Elliptical galaxies are more common in clusters of galaxies, which contain hundreds to thousands of galaxies, all bound together by gravity.

According to cold dark matter models, or hierarchical models, dark matter halos form from the clustering and merging of "small" dark matter halos.

Galaxy Trans-Formation

Galaxies evolve due to internal changes (for example, changes in color, brightness, chemical composition basically due to the star-formation history), and due to the interaction with other galaxies.

Collisions are also a factor that

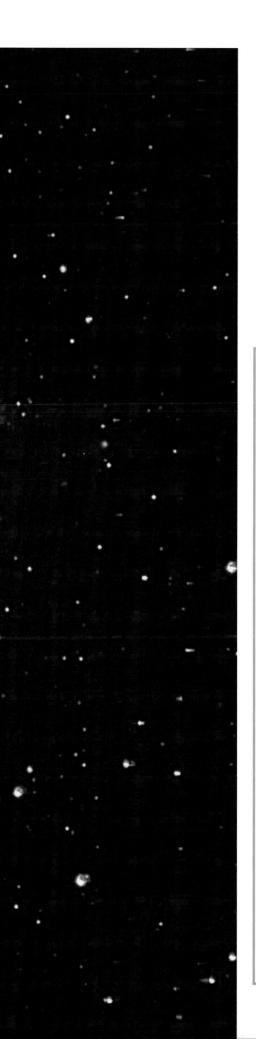

shapes galaxies. Of course, we cannot live long enough to see how galaxies merge. It would take several billion years!

So how can we check these ideas with observations? Astronomers are like detectives, following clues. Which signs can we find in galaxies that can lead us to think that there is, or was, a merger? These are some:

Images of pairs of galaxies may reveal tails and bridges of stars and gas that are sign of interactions.

Counter-rotation. In some galaxies that otherwise look pretty "normal," there is evidence that one of the components is counter-rotating or rotating orthogonally to the other component. For example, in a stellar disk, the inner disk could be rotating in the opposite direction of the outer disk, or the spheroidal component opposite to the disk component.

Structural details in elliptical galaxies. For instance, elliptical galaxies with dust lanes have undergone a major event at some point in their evolution. The younger population of stars in these galaxies could have formed at a later stage of the evolution of the galaxy through either a merger event or a secondary in situ star-formation burst by the acquisition of gas from the environment. NGC 5128 (also known as Centaurus A) is a prototype of this class of galaxies (for more information, see "NGC 5128: A Merger Tale").

Observations show that collisions trigger bursts of star formation. Computer simulations of such collisions confirm that the merger of two spiral galaxies can form an elliptical galaxy. ●

NGC 5128: A Merger Tale

One of the gems of the southern sky is, without a doubt, the galaxy NGC 5128 (also known as Centaurus A). This galaxy has inspired about a thousand scientific papers and still poses several questions.

There are many reasons that have made this elliptical galaxy particularly interesting to study.

NGC 5128 is the nearest giant elliptical galaxy to us; its distance is only 12 million light-years away.

It is also the nearest active galactic nucleus, hosting both a supermassive black hole of 100 million solar masses and a powerful radio source.

Furthermore, NGC 5128 is the nearest galaxy with shells; it contains a central dust lane, and boasts a bimodal populous system of 1,700 globular clusters and a hundred active star-forming regions along its dust lane.

NGC 5128 is part of a group of 25 galaxies, and there is indirect evidence to believe that it is the result of a merger of an elliptical galaxy and a small spiral galaxy. The "signs" that a major event occurred in the past are the following:

- A warped disk of dust, gas, and young stars
- Shell structures
- Bimodality of the globular cluster system.

The image was taken by José Funes, S.J., at Cerro Tololo Inter-American Observatory in Chile and processed by Sanae Akiyama. The stellar body of the galaxy is shown in blue and emission of the ionized gas in red.

Fr. José Funes, S.J. (Argentina), the director of the Vatican Observatory, specializes in the comparison of galaxy morphology, and he has organized two international meetings on galaxy morphology and evolution at the Vatican. This chapter was written specially for the book.

CLUSTERS OF GALAXIES:

• Fr. Alessandro OMIZZOLO •

FAMILIES OF FAMILIES OF STARS

History

The great French astronomer Charles Messier was the first to notice clusters of galaxies. In his famous catalog of sky objects, published in 1784, he included 103 nebulae, of which 30 were galaxies; and he noted that in the constellation of Virgo, there was a strange concentration of galaxies. Likewise, William Herschel, the British astronomer, noticed a concentration of galaxies in the region of the constellation Coma Berenices. But Herschel also noticed other clusters of galaxies, such as those in the constellations Leo, Ursa Major, and Hydra.

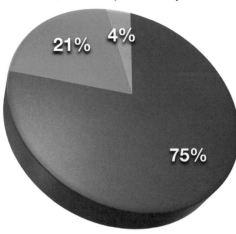

Left: The Coma Cluster of Galaxies, imaged by the Hubble Space Telescope. Credit: *NASA, ESA, and the Hubble Heritage Team.*
Above: The components of the visible universe: dark energy (blue), dark matter (green), and ordinary matter (orange).

Herschel's son John was also an astronomer; he studied the southern sky and cataloged over 6,000 nebulae. In 1888, Dreyer published his New General Catalog (which gave the familiar "NGC numbers" to the brightest nebula in the sky). From the beginning of the 20th century, many other catalogs have been published; all were derived from observations of galaxies. These catalogs have the signatures of the most famous astronomers of that time: Wolf, Curtis, Hubble, Shapley, Zwicky, Abell. A very important contribution is that of Abell, who, in 1958, started a new era in the investigation of clusters of galaxies with the publication of his catalog. Abell's work is important because he was the first to prove that there exist clusters of clusters (usually called superclusters), with dimensions in the range of about 200 million light-years; in 1962, he published the first list of these superclusters. In the following years, Abell and others published further such lists.

The Cosmological Point of View

The most reliable cosmological model today states that the geometry of the observable universe is a flat one. This means that the total density of the universe is close to the critical density necessary for the universe to be closed. Ordinary baryonic matter, that which makes up everything we know and our instruments can measure, apparently makes up only about 4 percent of the mass of the universe. The two main components of the universe seem to be in the form of non-baryonic dark matter and dark energy.

Non-baryonic dark matter is material that exerts a gravitational attraction, which is how we know it is there; it is responsible for the formation of cosmic structures. The even more mysterious dark energy has been deduced

Top: The Bullet cluster in optical wavelengths, as imaged by the Hubble Space Telescope. The two apparent groups of galaxies to the left and right are actually the same group, gravitationally lensed by a galaxy cluster in the center of the field of view.
Middle: To this image, the blue clouds are added, based on X-ray data from the Chandra Space Telescope.
Bottom: The red region is the calculated mass distribution in the cluster, based on the gravitational lensing seen in the top image. Note that the center of gravity is significantly offset from the center of the two clusters visible in x-rays. From this, one can conclude that the bulk of the mass doing the lensing is not visible, but dark, matter.
Credit: X-ray, *NASA/CXC/CfA/M. Markevitch et al.;* Optical: *NASA/STScI; Magellan/U. Arizona/ D. Clowe et al.;* Lensing Map: *NASA/STScI; ESO WFI; Magellan/U. Arizona/D. Clowe et al.*

to account for the pressure needed to explain the acceleration in the expansion of the universe that we are observing. The average quantity of the baryonic matter is about 4 percent of the total amount of matter. Indeed, baryonic matter is observable in space only because the gravitational force of the non-baryonic matter created "potential holes" and then moved the baryonic gas into such holes, where some of the baryonic matter could condense to form stars.

As we can see, normal matter is a very small part of the whole; it is to this part that galaxy clusters belong. They are self-gravitating systems with masses in the range $10^{14} - 10^{15}$ times the mass of our Sun, and with sizes in the range of 3 million to 9 million light-years. About 80 percent of this mass is dark matter; most of the rest is a diffuse intracluster plasma (ICM). Only a few percent of the mass of a galaxy cluster is made up of stars, dust, and cold gas mostly locked within the galaxies. All of these components are in a dynamical equilibrium inside the cluster's gravitational potential well. Nevertheless, the spatial inhomogeneity of the thermal and non-thermal emission of the ICM, and the kinematic and morphological segregation of galaxies, are a clear indication that processes of a non-gravitational nature are going on: merging and interactions between clusters. Furthermore, we notice that the fraction of clusters

that show these features grows with the redshift of the cluster — i.e., with the distance from us, which is to say, the distance from us in time.

According to the current "bottom-up" scenario, galaxies form first and then they collect to form a cluster of galaxies. In this model, the clusters are the most massive nodes of the filamentary large-scale structure of the cosmic web and form by anisotropic and episodic accretions of mass. This is in good agreement with most of the observational evidence. In this model, the universe is dominated by cold dark matter. Most of the baryons (ordinary matter) are expected to be found in a diffuse component and not in stars or in galaxies; about 50 percent of this diffuse component permeates the filamentary distribution of the dark matter.

Clusters of galaxies are a particularly rich source of information about the underlying cosmological model and make possible the realization of some critical tests that can help to distinguish between the various models. Clusters are the largest and most recent gravitationally relaxed objects to form, because their structure grows hierarchically under the action of gravity from the smallest structures to the largest ones. Clusters form this way, and so their growth and development directly trace the process of structure formation in the universe.

Observing Clusters with Optical Telescopes

All of the stars in all of a cluster's galaxies represent only a small fraction of the total mass of the cluster. In fact, a large amount of the cluster's mass is in the form of a hot gas emitting a lot of energy in X-rays and microwaves, and thus clusters are strong emitters in these two regions of the electromagnetic spectrum.

As we said, Messier and Herschel noticed an unusual concentration of galaxies in the constellations of Virgo and Coma Berenices, but they did not interpret these concentrations as a physical connection between the galaxies they observed. Since that very early time, the discoveries quickly grew as the power of the telescopes in use grew,

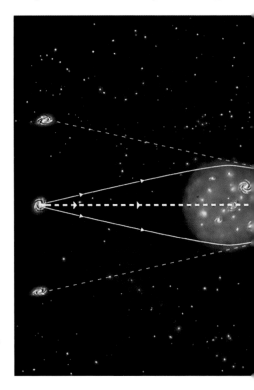

leading ultimately to the publication of the catalogs of clusters that Abell and his collaborators published in 1958 and 1989.

We know that luminosity is an indication of the mass of the universe; in the same way, total optical luminosity gives a first indication of the mass of the cluster. To do this, we use the so-called luminosity distribution function of the cluster because this function is similar from cluster to cluster. Thus, through this function we can estimate the cluster's mass. Abell used this function to define a new parameter, known as a cluster's richness. With this parameter, it is possible to catalog the clusters according to the number of galaxies brighter than a given limiting magnitude. In this way, when we speak of the number of galaxies in a cluster, we can mean the mass of the cluster, or at least the visible mass. So we can classify clusters according to their richness class with values that, in Abell's scale, range from 0 to 5: class 0 clusters contain fewer galaxies than the minimum defined by Abell, while 5 indicates the richest and most massive clusters.

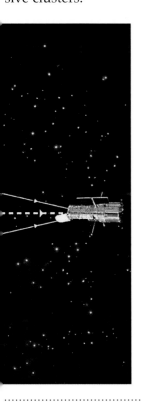

The identification of these clusters visible in the optical wavelengths was not as easy as we might suspect, owing to projection effects: Are two galaxies that appear next to each other in our telescope really close to each other, or are they merely by chance in the same line of sight as seen from Earth, with one of them much closer to us than the other? That is why Abell decided to introduce detailed criteria through which one can say if a given ensemble of galaxies do form a cluster, and which galaxies belong to the cluster.

These criteria are particularly useful when one has to deal with clusters that are at larger distances. In fact, for these it is not so easy to distinguish the cluster's borders, owing to the distance from the observer and to the contrast of clusters against the background galaxy counts, which decrease with cluster distance. We can solve this problem, at least partly, by taking into account the colors of the galaxies: galaxy colors can help identify distant clusters because many cluster galaxies are redder than galaxies of the field with similar redshift, owing to their lack of ongoing star formation.

The Discovery of Dark Matter

Another way to determine if a galaxy belongs or not to a given cluster is to measure the relative velocities of the individu-

al galaxies (which can be determined by studying the red- and blue-shift of spectral lines emitted by a given galaxy, relative to the average redshift of the entire cluster). It seems reasonable to think that the range of the velocity of the galaxies inside a cluster follows the typical statistical pattern known as a Gaussian distribution; if so, then it is very unlikely that galaxies whose velocity falls outside the limits of this Gaussian distribution would be members of the cluster, and so they are rejected as cluster members.

But the velocity data can give other important information about the physical properties of the cluster. In fact, the measured values of these speeds soon raised an important problem that can affect the cosmological model. The measured speed values are correct only if the cluster has much more mass than can be accounted for from the light-emitting matter such as stars and nebulae. The first one to raise this problem was Zwicky in the 1930s, and he provided the first evidence for the existence of a new kind of matter called dark matter. What Zwicky derived for the Coma cluster, Smith obtained a few years later for the Virgo cluster.

There is another way to estimate the mass of a galaxy cluster, using a physical phenomenon called "gravitational lensing."

Light rays emitted from a source located behind a galaxy cluster are bent by the mass of the galaxy cluster. They can form multiple images of the same object in different places around the cluster, so that an object (usually a galaxy) otherwise not visible becomes an observable one. The amount of the deflection of the light depends on the cluster mass. So, if we measure the deflection angle, we have also indirectly measured the mass that caused the observed deflection.

Again, the mass we derive is larger than that obtained by simply adding the masses of the single galaxies forming the cluster. In this way, we have an indication not only of the value of the actual mass of the cluster, but also proof of the existence of a kind of mass that we otherwise would fail to observe.

Observing Clusters with X-rays

Clusters of galaxies are also strong X-ray emitters because of the low efficiency of the galaxy formation process. In fact, only about 10 percent of the baryons in the universe are located between the stars in the galaxies, while the remaining 90 percent go adrift in intergalactic space. This intergalactic baryonic gas is compressed inside the deep potential wells of galaxy clusters and heated up to such high temperatures that the gas begins to emit X-rays. (The mechanism of emission of the X radiation is the thermal bremsstrahlung of charged particles when the particles undergo a strong deceleration.) So the temperature we get from the X-ray emission is a measure of the strength with which the gas is compressed by the cluster, while the intensity of the emission lines in the X band indicates the chemical abundances of elements such iron, oxygen, and silicon in the space between clusters (ICM). This extended X-ray emission from galaxy clusters was first observed at the beginning of the 1970s.

There is another relevant property of the clusters that we must consider. Clusters in hydrostatic equilibrium have a plasma temperature (plasma is totally ionized matter) that is strongly related to the overall matter. So if we succeed in measuring this temperature, we could get more information about the overall mass of the cluster. The problem at this point is the number of X-ray photons we collect from the cluster: the greater this number, the more information we get!

In fact, for iron, oxygen, and silicon, it is rather easy to measure their abundances from the energy emitted in the form of lines in the X-ray spectra. This emission is done through collisional excitation: every time there is such a collision, a photon is produced and leaves the cluster. What we find is that the iron abundance at the center of the cluster is higher, especially when there is a giant galaxy in the center of the cluster. This growth in the abundance reflects also the presence in that region of phenomena connected to the last stages of stellar evolution such as the explosion of supernovae, neutron stars, and white dwarfs.

Current research is focused on learning if there is some correlation between the optical and X properties of

Right: The galaxy cluster Abell 520. This is a composite of three images produced in the same way as the image on page 103: optical light obtained by the Hubble Space Telescope, X-rays imaged by the Chandra X-ray Telescope (shown in blue), and a map of the mass of the cluster as determined by how it lenses more distant galaxies. Credit: *NASA/ ESO, Space Telescope and Chandra observatories.*

the galaxy clusters. This is still a wide-open field, where we are still looking for more insights. As an example, it is interesting to see how important the contribution of the X-ray imaging is to understand the morphology of a galaxy cluster. On page 103, we show images of a given cluster seen in various bands of the electromagnetic spectrum.

Deducing Large-Scale Structure of the Universe

Even from the earliest observations of galaxy clusters, it was known that clusters are not casually distributed in the universe, but show a tendency to form large concentrations called "superclusters." The connection between clusters and the large-scale structures of the universe became clear with modern large surveys of galaxies. This connection appears also in the X-ray waveband: the X-ray luminosity of clusters and the irregularity of the X-ray surface brightness maps are larger in environments with a higher cluster number density.

The masses of galaxy clusters, derived with the technique we previously spoke of, vary between 10^{14} and 10^{15} solar masses and can be split this way: about 80-90 percent of the overall mass of the cluster is dark matter (emitting no detectable radiation), the ICM contributes about 10-20 percent, and the galaxies the remaining fraction. When one considers cosmological models in which mass density is dominated by non-baryonic dark matter, it is rather easy to explain the clusters' properties and their evolution with redshift.

A large fraction of clusters shows the existence of substructures, both in the distribution of galaxies and in the morphology of their X-ray emission. Moreover, the morphology in the X-ray band becomes more irregular with increasing redshift. This suggests that clusters accrete matter from the surrounding regions. A clear example is the "Bullet Cluster" (p. 103). The X-ray morphology is increasingly irregular with increasing redshift. Accretion of matter is a natural explanation for the morphology of the X-ray emission and metallicity maps; this accretion can also be responsible for the turbulence in the ICM, which can generate high-energy cosmic rays.

It is also possible to observe the

hot gas in clusters through its effects on the cosmic microwave background. Soon after the discovery of this background radiation in the mid-1960s, Weymann studied the effects on this radiation of a physical process known as Compton diffusion, a shift toward slightly higher energies of some of the background radiation photons when they pass through the hot intergalactic gas. In 1970, Sunyaev and Zeldovich predicted that the hot gas of the clusters could produce this distortion of the spectrum, an effect now known as Sunyaev-Zeldovich (SZ) Effect.

It is useful to compare the X-ray and SZ properties of a cluster. The X-ray and SZ observations of a cluster are complementary because they give not only the gas temperature but also the integral of the gas density along the line of sight through a cluster. Combining this data, it is possible to study the outer regions of a cluster where the X-ray

surface brightness is difficult to measure while the SZ signal remains strong.

Other applications of this effect concern the study of the distance and angular dimensions of the cluster, which are strongly connected with the scale and geometry of the universe. So, once again, galaxy clusters are proving to be an important instrument to study also the large-scale structure of the universe. ●

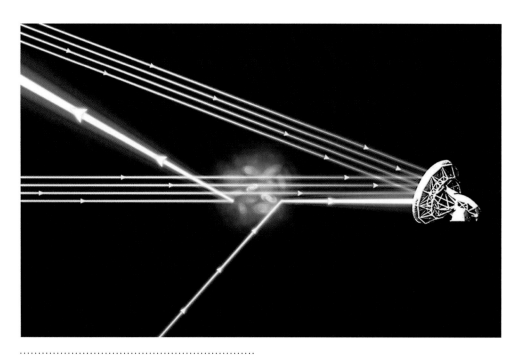

Above: In the Sunyaev-Zeldovich (SZ) effect, photons in the cosmic microwave background (CMB) interact with electrons in the hot gas that pervades the enormous galaxy clusters. The photons acquire energy from this interaction, which distorts the signal from the microwave background in the direction of the clusters. The magnitude of this distortion depends on the density and temperature of the hot electrons and the physical size of the cluster.

Fr. Alessandro Omizzolo (Italy), a priest of the Diocese of Padua, divides his time between Castel Gandolfo and the Padua Observatory, Italy. He has worked on quasars, active galactic nuclei, and galaxy clusters. This chapter was written specially for this book.

SEEING STARS

Photo by Ron Dantowitz, Clay Center Observatory.

THE SPECOLA IN CASTEL GANDOLFO

Since Pope Pius XI moved the observatory (Specola Vaticana) from the walls of the Vatican in 1935, its headquarters have been situated on the palace and in the villas of the Pope's summer residence in Castel Gandolfo, a small Italian village 30 kilometers southeast of Rome.

Here, perched on the rim of an extinct volcano, the telescopes of the Vatican Observatory had an unimpeded view of the skies … until, inevitably, city lights encroached on the area.

But even as the telescopes fell out of regular use (they are still available for special astronomical events), the location has remained an active site of the observatory's work.

There is something about the wonderful views — Lake Albano and Monte Cavo to the east, and the plains south of Rome leading to the Mediterranean to the west — that makes it an attractive venue for small meetings, summer schools, and simply the day-to-day work of thinking, modeling, and writing about the universe as seen through the lenses of astronomy.

Approaching Castel Gandolfo

rriving from Rome, the first view a visitor gets of the Vatican Observatory are the two telescope domes perched on top of the Papal Palace itself. Sitting atop the pale golden mass of the palace, itself located on a high point of land, they command the view from every direction. And in their form they echo the larger dome, designed by Bernini, atop

the church of St. Thomas of Villanova in the nearby town square; and they echo the vault of the sky under which church, telescopes, village, and lake all lie.

Within the village itself are many wonderful cafés, restaurants, and souvenir stands for all the visitors who come to see the Pope's home (and to see the Pope himself, when he is in residence during the summer). One can also find all the regular shops that supply the needs of the villagers: grocers, butchers, and fruit sellers; newspaper stands and hardware stores.

And, of course, pizza and gelato!

Photo credit: *Tijl Kindt*

The Telescopes
at Castel Gandolfo

Four telescopes can be found in the domes on or near the Papal Palace.

Atop the palace itself, on a dome built over the elliptical staircase designed by Bernini (*below*), is the 40 cm aperture f/15 Zeiss refractor (known as the *Visuale*) installed in 1935. It is still the workhorse of the observatory here; it has excellent optics and, with the invention of lightweight CCD cameras, it is now able to produce images that could never have been imagined when it was installed.

Nearby, in a dome overlooking the lake and bearing the motto *Deum Creatorem Venite Adoremus* given to the observatory by Pope Pius XI, the twin telescopes of the Zeiss Double Astrograph can be found (*right*). Though the astrocamera half is no longer is use (it was designed to use photographic plates which are no longer available), the 60 cm Cassegrain reflector (*mirror, opposite top*) still functions well. It has been used in recent years to record transits of stars behind Pluto and Saturn's moon Titan, and mutual events of Jupiter's

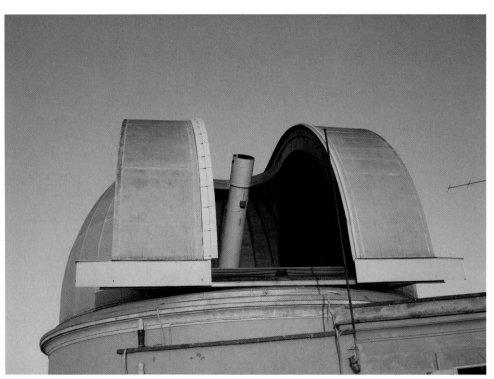

DEVM CREATOREM
VENITE ADOREMVS

moons. Both telescopes observed the impact of comet Shoemaker-Levy 9 onto Jupiter in 1994; because of their location, they were among the first in the world to record the black spots on Jupiter that resulted from this impact.

In the Papal gardens nearby, the 1891 *Carte du Ciel* telescope (*opposite, bottom*) and the 1958 Schmidt camera (see p. 76) can be found. Due to increased light pollution as the city has grown up around Castel Gandolfo, neither telescope is in operation today.

Photo credits: *Ron Dantowitz; Specola archives*

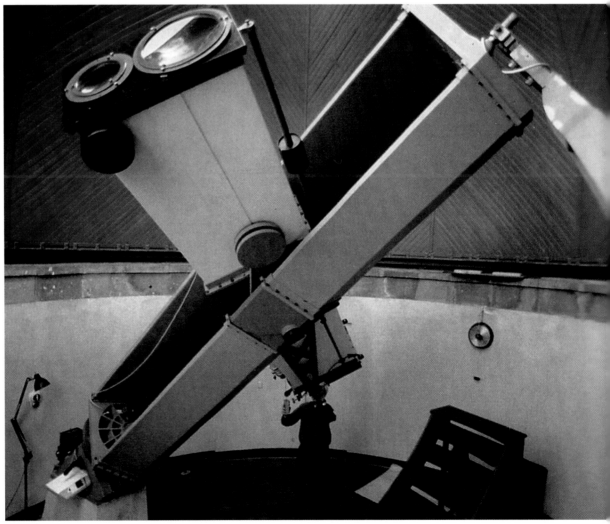

Observing the Transit of Venus

On June 8, 2004, a twice-in-a-century astronomical event occurred which could be seen from Castel Gandolfo: the passage of the planet Venus in front of the Sun. In past centuries (the last such transit had occurred in 1882), these "transits of Venus" were used to determine the scale of the Solar System. Today, they are mostly an astronomical curiosity, but nonetheless a beautiful sight to behold.

At the Papal Palace, a group of 50 amateur and professional astronomers, organized by *Sky and Telescope* magazine, set up their telescopes for a clear and remarkably steady view as the Sun rose over Lake Albano. (Images from this transit can be seen on pages 32 and 137.)

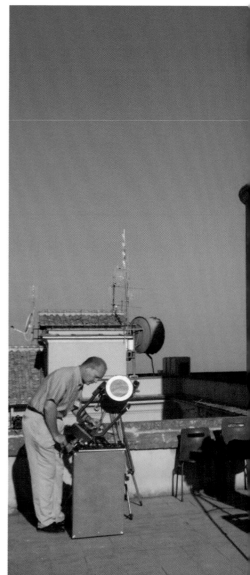

Photo credits: *Fr. Chris Corbally, S.J.*

VOSS SUMMER SCHOOL TOPICS

Year	Topic
1986	*The Structure and Dynamics of Galaxies*
1988	*Star Formation in Galaxies*
1990	*The Structure of Galaxies and the Spectral Classification of Stars*
1993	*The Nuclei of Galaxies*
1995	*Theoretical and Observational Cosmology*
1997	*Comets, Asteroids, and Meteorites*
1999	*Single Stars and Close Binary Systems*
2001	*Stellar Remnants*
2003	*Galaxy Evolution*
2005	*Astrobiology*
2007	*Extrasolar Planets and Brown Dwarfs*

Vatican Observatory Summer Schools (The VOSS)

In 1985, Fr. Martin McCarthy, S.J., realized that because the Vatican Observatory does not give degrees, we had little chance for contact with young people. That was a shame, he thought, since they are not only the promise for the future of the field, but also often among the most active and imaginative researchers even now. How could this be solved? He hit upon the idea of a regular month-long summer school to be held at the observatory at Castel Gandolfo on some special topic in astronomy and astrophysics. A year later, the Vatican Observatory Summer School program was born. Since then, the observatory has held such a school roughly every two years. Eleven in all

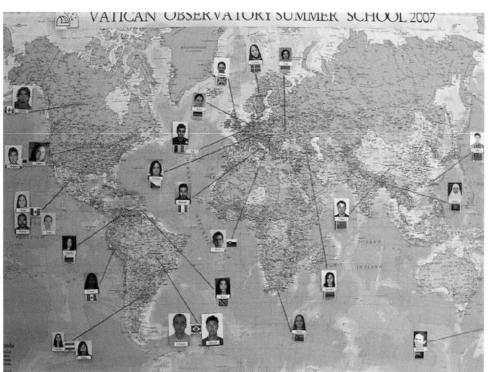

have been held to date.

The schools are open to advanced undergraduates and beginning graduate students in astronomy from around the world; more than 55 nations have been represented so far. The only rules limiting who can attend are that the students must show evidence of likely success as professional astronomers, and no more than two students from any given nation are accepted. The great challenge is choosing 25 students out of an applicant pool of more than 200 candidates. A majority of the students selected come from underdeveloped countries. Tuition is free, and significant financial support ensures that every student accepted is able to attend.

To date, nearly 300 young astronomers have passed through these schools, and over 85 percent are still active in astronomy. They have gone on to work

at the most prestigious institutions around the world, like the Max Planck *Institut für Astrophysik*, the European Southern Observatory in Chile, and leading astronomy programs at universities in the United States, including Arizona, Caltech, and Yale.

The faculty are among the most notable astronomers of their age, drawn from leading observatories and universities around the world. Among them have been Vera Rubin, winner of the 2002 Gruber Cosmology Prize; Frank Shu, later president of the National Tsing Hua University in Taiwan; and Michael A'Hearn, the principal investigator of the NASA Deep Impact mission to Comet 9P/Tempel. The most recent school was organized by an alumnus of the 1986 VOSS, Dante Minniti.

The 11th Vatican Observatory Summer School was held from June 9 to July 6, 2007. The topic that year was the search and study of Extrasolar Planets and Brown Dwarfs: bodies, most of them only discovered in the past 10 years, that provide a link between stars and our own Solar System. (To learn more about them, see Fr. Koch's chapter in this book beginning on page 152.) As well as being systems that hold the tantalizing promise of harboring planets like Earth, studying these bodies helps us understand how our own Solar System was formed.

Twenty-six young astronomers came from around the world, from Argentina to Sweden, Indonesia to the United States, representing every continent on Earth. (One student, from New Zealand, had even done research in Antarctica before she came to the school!) In addition, young Jesuit astronomers from Mexico and the Czech Republic also took part in the program. Fifteen of the students came from less-developed nations; 16 of the students were women.

In addition to the daily lectures from some of the leading experts in the world on these topics, the students had the chance to explore Italy. A weekend trip to Florence featured a visit to Galileo's home, while day trips from Castel Gandolfo ranged from tours of historical abbeys to a day at the beach. And of course, Rome itself was a popular afternoon and weekend destination for the students. Indeed, the school began at the Vatican itself, in a private audience with Pope Benedict XVI — which got front-page coverage in the Vatican newspaper *L'Osservatore Romano*! (His remarks to the students can be found on page 212.)

Vatican Observatory director Dr. José Funes, S.J., served as dean of the school, while Dr. Dante Minniti of the Pontificia Universidad Catolica, Santiago, Chile, organized the academic program. Among the lecturers at the school were Dr. France Allard, Centre de Recherche Astrophysique de Lyon, Lyon, France; Fernando Comeron, from the European Southern Observatory (ESO); Garching bei München, Germany; and Didier Queloz, Geneva Observatory, Sauverny, Switzerland.

They were joined by a number of guest lecturers from Europe and America, among them the president of the International Astronomical Union, Dr. Catherine J. Cesarsky of ESO.

The photographs here are from the 2007 school, courtesy of Tijl Kindt, one of the VOSS students.

Home

Castel Gandolfo is more than where we work; it is also our home. This is where we live the seasons of the year, and the seasons of our lives.

Whether it is celebrating a Mass with our friends and coworkers, or taking a walk past the snow-covered Christmas tree in the piazza, we are a part of a movement as regular, and as beautiful, as the movements of the Sun and stars and planets.

We are privileged to be here ... not only because we are guests of His Holiness, not just because this setting is beautiful. We are privileged to be alive in this wonderful universe, and to have been given the talents and the opportunities to get to learn more about it, and the One who created it. ●

Above and left: Mass at the retirement of Fr. George Coyne as director of the observatory; concelebrating with him are Sabino Maffeo, longtime administrator of the observatory in Castel Gandolfo and rector of the Jesuit Community during much of Fr. Coyne's tenure, and Fr. José Funes, the incoming director of the observatory.

Top: The Double Astrograph dome in winter.
Right: In 2009, the main offices and living quarters of the Vatican Observatory will be transferred to this historic monastery in the Papal gardens adjacent to the Papal Palace, completely refurbished to support the next century of the observatory's work.

AT THE VATT

PHOTOS BY
• Fr. RICHARD BOYLE, S.J., AND A.G.D. PHILIP •

On a mountaintop more than a hundred miles east of Tucson, Arizona, the Mt. Graham International Observatory (MGIO) looks through clear, dry desert skies to the wonders of the universe. One of the instruments on that mountain is the Vatican Advanced Technology Telescope — known more simply as *The VATT*.

What is a typical "observing run" like for the astronomers who work at the VATT? We follow here the Vatican astronomer Fr. Rich Boyle and his colleagues Dave Philip, Union College, New York; Luisa Zambrano, Universidad Metropolitana, Puerto Rico; and Olga I. Pintado, Universidad Nacional Tucuman, Argentina. In late spring, they prepare for a weeklong stay at the telescope to survey stellar populations in the Milky Way by taking images of selected star fields using the "Strom-Vil" filters. These filtered images will allow them to classify all the stars in their field of view, and help lead to a better understanding of how different types of stars are distributed through the Galaxy.

Right: *On the horizon is Mt. Graham, as seen from Mt. Lemmon, the site of another set of telescopes north of Tucson also operated by the University of Arizona. Note the white patch of snow just visible on the distant mountaintop.*

The first challenge is simply getting to the mountain itself; it is a two-hour drive through the desert to the base camp, located outside of Safford, Arizona. A popular stop en route is to have a picnic lunch in a particularly scenic area known as Texas Canyon.

Before heading up the mountain, all observers must check in at the base camp, the local headquarters for the MGIO, to obtain the necessary keys and permits for working on the mountaintop. The land on which the telescopes are situated is a delicate ecosystem, managed by the National Forest Service, and access is restricted. (Visitors can see the telescopes on tours arranged during the warmer months at the Discovery Park museum in Safford.)

Above and top: *The group stops at Texas Canyon for lunch en route to Mt. Graham.*

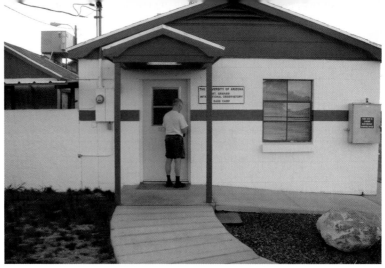

Above: Mt. Graham as seen from the base camp.
Top: The base camp outside Safford.

After stopping for permits at the base camp, there still remains more than an hour's drive up the mountain to the observatory. Much of the way is paved, going past summer homes and campgrounds, but the final nine miles is a rough dirt road. And even the paved road is slow going; tight twists and steep climbs limit a vehicle's speed to less than 15 miles per hour (25 kilometers per hour) in many places.

Traveling up the mountain is like traveling north; as the air gets thinner, it also gets colder. Along the way you can see the vegetation change from the semi-arid Arizona desert to pine forests

at the top of the mountain. By the side of the road, dozens of small waterfalls are fed by the melting snow farther up the mountain. By the time you reach 10,000 feet, the climate is similar to Canada.

But this is still Arizona, and water can be scarce. Thus, the mountaintop is vulnerable to forest fires. Twice in the last 15 years, fires started by lightning have threatened the observatory. The damaged trees on the last stretch of road leading to the telescopes are a stark reminder of these past fires, and of the many firefighters who protected the telescopes.

SEEING
STARS
AT THE VATT

At the top of the mountain are the three telescopes of the Mt. Graham International Observatory.

Above: This view, taken from the Large Binocular Telescope, shows the VATT (left)
and the 10-meter dish (right) of the Sub-Millimeter Telescope (SMT).
Top: The VATT building, behind Luisa, houses the 1.8-meter Alice P. Lennon Telescope
and the Bannan Astrophysics Facility, providing technical support for the telescope,
along with four bedrooms and a kitchen for the astronomers.

126

Top: *The dome of the VATT peers out from the trees surrounding the observatory. Note also the microwave tower, which provides telephone and high-speed Internet links for the observers.*

Middle and lower: *From the dome of the VATT, one has an unobstructed view of the Arizona sky in all directions.*

A snowy scene greets the observers when they rise. For Luisa, raised in Colombia and Puerto Rico, seeing snow is quite a novelty. But even a New Yorker like Dave is amused at the way the sticky spring snow will sometimes roll itself into a donut shape as it melts and starts to slide downhill.

Life on the mountaintop has its own rhythms. After working all night, the observers generally try to sleep in as long as possible into the afternoon. Late afternoons, they can visit the other telescopes on the mountaintop, and prepare their evening meal before the night's work of observing begins.

The VATT residence boasts a full kitchen on the ground floor, downstairs from the telescope control room. The observers are responsible for preparing their own meals. (And midnight snacks!)

At sunset, the work begins. It's time to open up the dome and power-up the telescopes.

When the domes are opened, the telescopes are pointed to the east *(below)*, away from the direction of the setting Sun. This is to prevent the last rays from heating up the inside of the dome or striking the delicate cameras.

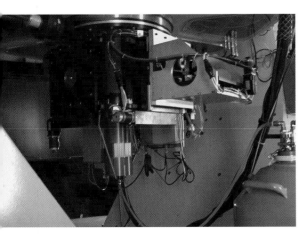

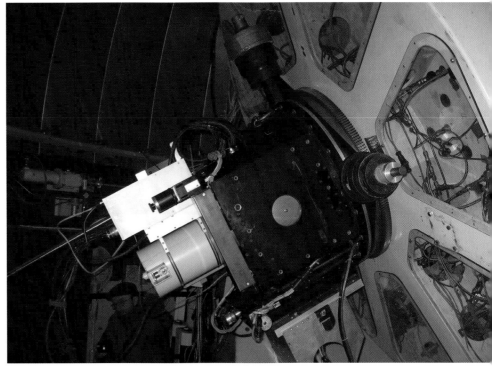

A telescope like the VATT requires a lot of steps to get ready for observing. First, the dome must be opened and the mirror uncovered *(top, right)*.

Vents are opened and fans turned on to cool the dome; and a special system brings coolant to the telescope mirror, ensuring that it is maintained at precisely the same temperature as the ambient air to prevent currents that would distort the telescope's images. (Some of the piping can be seen on the back of the mirror, *above.*) The camera shutter itself may need adjusting *(top left)*. And every night, liquid nitrogen is pumped into the electronic camera to cool the CCD chip, making it especially sensitive to even the faintest levels of light.

This job, which involves crawling briefly underneath the telescope, is usually given to the youngest member *(opposite top)* of the observing team!

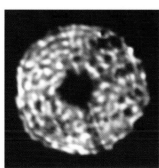

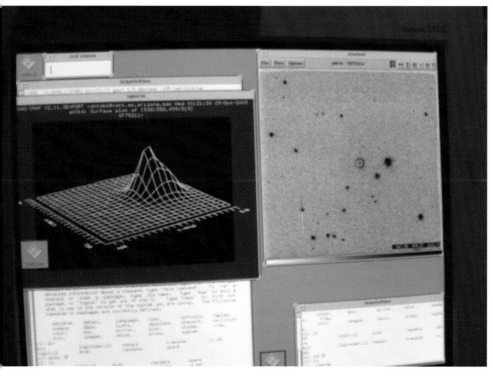

tors display images from TV cameras around the observatory; one is pointed at the night sky so that the observers in the room can be warned if clouds are coming or going, while another camera resides in the dome, where light can be turned on and the telescope watched as it is steered from one object to another in the sky. Other monitors track the CCD camera in the telescope's finderscope, and keep track of the weather station, indicating local conditions like winds or humidity that might affect the performance of the telescope.

One of the first tasks for a night's observing is to point the telescope at a star and throw the image out of focus (*above left*). In the resulting fuzzy image, the secondary mirror (and the struts that support it) are visible as a dark spot. By tilting ever so slightly the position of the secondary mirror, the optics can be perfectly aligned and collimated; when that occurs, the black spot is exactly in the middle of the "donut."

And, of course, the computers are also used to monitor the results coming from the science camera itself. Using special software tools (*left*), the signal from each star can be studied, and the amount of light received by the CCD camera recorded.

Then the work of observing begins. The astronomers never go into the dome itself while taking an image; the heat of their bodies would stir up the air too much. Instead, all observing is done by computer from a control room, well removed from the dome. All images are obtained by a computer-controlled CCD camera; the VATT doesn't have any eyepieces! Likewise, all control of the telescope's motion is done by computer.

Rich (*opposite top*) and Dave and Olga (*opposite left*) track the motion of the telescope and the images obtained by the science camera. Other moni-

After a week, it's time to come down. The trip down the mountain is bittersweet. After an observing run, the top of the mountain has begun to feel like home, and it can be hard to leave. In addition, the astronomers know that — if they were lucky with the weather! — they have months of work waiting for them analyzing the gigabytes of data on their computer disks, before they are ready to go back to the telescope again.

The camaraderie of the observing group is another feature of life on the mountaintop that will be missed. You become great friends over the course of a week, chatting about family and friends, as well as about astronomy, while waiting for the images to come from the camera.

And the quiet setting, the peacefulness of working through the night, and the glory of both the mountaintop and the stars are more than a little reminis-

cent of life in a monastery. You do feel removed from the daily bustle of life back in the city, and a little closer to Creation.

And if you're lucky, the bad weather you were afraid might spoil your observing run will wait until it's time to leave the mountain. ●

A STAR'S COMPANIONS

Above: The planet Venus crosses the face of the Sun, as seen from Earth, twice a century. This image was taken from outside the dome of the Double Astrograph in Castel Gandolfo on the morning of June 8, 2004, by Ron Dantowitz of the Clay Center Observatory. He used a 1/20,000th of a second exposure with a Meade 7" Maksutov with a full aperture 7" energy rejection filter and a 0.5a DayStar University filter. Visible are the granulation patterns on the surface of the Sun and the thin ring of atmosphere surrounding the disk of Venus.

SUN AND PLANETS, PAST AND FUTURE

• Br. Guy CONSOLMAGNO, S.J. •

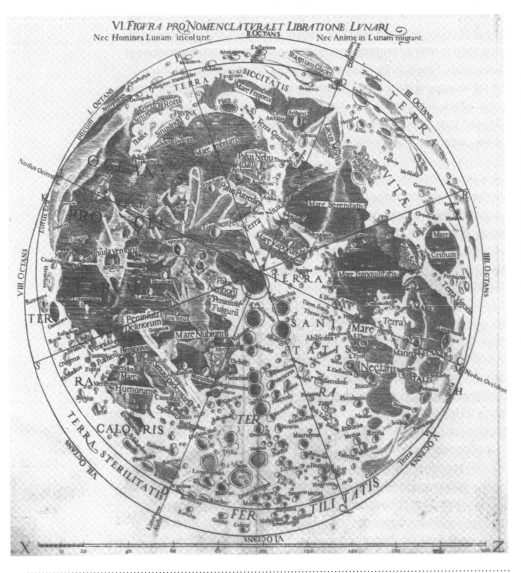

VI. FIGVRA PRO NOMENCLATVRAE ET LIBRATIONE LVNARI

Nec Homines Lunam incolunt.   B. OCTANS   Nec Animæ in Lunam migrant.

Above: Francesco Grimaldi and Giovanni Battista Riccioli published this map of the Moon in 1652, which established the nomenclature used to this day. Controversial even today is the significance that, less than 20 years after Galileo's trial, they named the most prominent crater for Copernicus. Other prominent supporters of the heliocentric idea, including Aristarchus and Kepler, also have craters named for them near Copernicus; likewise, not far away, are the craters Grimaldi and Riccioli.

Stars are the most visible members of the universe. It is by their light that all other objects can be seen. The light of our own star, the Sun, lets us understand in a close-up way how stars in general behave. But its radiant beams reflected off the surface of nearby lumps of rock, ice, and gas have also brought to light these other worlds orbiting around it, fascinating places in both their similarities and differences with our own planet, Earth. And over the years, astronomers at the Vatican and their collaborators have contributed significantly to our understanding of this Solar System.

A mere 20 years after Galileo's trial, Jesuit astronomers at the Roman College, Francesco Grimaldi and Giovanni Battista Riccioli, produced a map of the Moon that was the most accurate of its day and, more importantly, created the nomenclature for lunar features that is still in use. Other 17th-century Jesuit planetary astronomers included Gilles-François de Gottignies, who observed the comets of 1664, 1665, and 1668, and Athanasius Kircher, who made some of the first detailed telescopic drawings of Jupiter and Saturn.

In the 18th century, the polymath Jesuit Ruggiero (Roger) Boscovich's wide-ranging activities included the study of transits, cometary orbits, and the optics of telescopes. Boscovich organized the Jesuits' astronomical observatory in Brera, outside Milan, and at one time he got approval to build an obser-

triangle"), which "forms a sort of large *channel (canale)* between two red-colored *continents:* in another figure here there is another sky-blue channel that connects two spots of a darker shade than the previous ones."

At that time, the dark markings themselves were still poorly understood. "Are these spots seas, clouds, or continents?" he asked. "Is there an atmosphere on Mars?" He noted that the larger dark features had been seen

vatory at the Roman College on the roof of the nearby St. Ignatius Church. However, it was a hundred years later before another Jesuit, Angelo Secchi, could complete that project.

From 1803 to 1824, Vatican astronomer and priest Giuseppe Calandrelli and his collaborators Andrea Conti and (after 1816) Giacomo Ricchebach produced eight volumes of *Opuscoli Astronomici (Astronomical Tracts)*, detailing their research on the Sun, planets, comets, and stellar occultations. The Jesuit priests Etienne Dumouchel and Francesco de Vico were the first to recover Comet Halley in 1835, and De Vico also determined the orbits of the Saturnian satellites Mimas and Enceladus.

Mars

The scope of Mars activities over 150 years of Papal observatories ranges from the first hints of *canali* to actual robot spacecraft landing of the surface of that planet.

A hundred and fifty years ago, Fr. Angelo Secchi published a book on the Solar System, *Quadro Fisico del Sistema Solare (A Physical Outline of the Solar System)*, which included his description of Mars as seen through his telescope, finally built on the roof of St. Ignatius Church. In that book, he described observing what is now called Syrtis Major ("a kind of large, beautiful azure

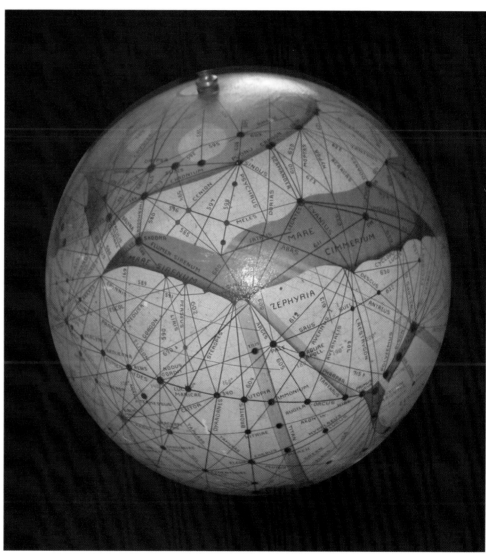

Top: This image of Mars was taken by the expert amateur astronomer Claudio Costa with a small color webcam inserted in place of the eyepiece on the Vatican Observatory's 40-cm Zeiss refractor on the night of September 9, 2003. As is typical for refractor telescopes, south is up. Note that, while Lowell's canals are not visible in the telescope, the dark channels first noted by Secchi are clearly visible.
Above: The globe of Mars, hand-painted around 1916 by Ingeborg Bruhn, is based on the maps of Percival Lowell. As in a telescope, south is up. The orientation here matches that of Mars seen in the previous image.

for many years by different observers, but concluded: "If in our complete ignorance we might be allowed to propose a hypothesis, it is not improbable that the large azure channels (*i grandi canali azzurri*) are seas, and the red parts continents, and the white spots [at the poles] masses of snow or clouds, since Mars for all its distance from the sun is not all that different than the Earth … if it has an atmosphere, as is probable considering how the light from the limb is always much darker than that from the center of the disk, it certainly should be much more tranquil and transparent than that of Jupiter or the Earth."

This was the first application of the Italian term *canali* to describe certain dark features on the surface of the planet. The features he saw were real; Secchi's use of *canali* is clearly different from how the Italian astronomer Giovanni Schiaparelli used the word some 20 years later. From the observatory founded a hundred years earlier by Boscovich in Brera, Schiaparelli mapped Martian albedo features and defined the cartography and nomenclature still used today for Martian surface features. But he specifically charted a network of thin lines connecting the dark regions. Published as a series of papers in the *Acts of the Royal Lincean Academy* in 1877-1878, his work noted that the existence of these *canali* was controversial, but found support in the fact that previous observers (mentioning Secchi among others) had also reported them.

This terminology was famously adapted by the American astronomer Percival Lowell, who translated *canali* as "canals" and inferred an intelligent origin for them — an interpretation that Schiaparelli never specifically repudiated. Thus, though the term *canali* actually originated with Secchi for the dark regions of Mars, whose presence is well con-

Above: Fr. Roger Boscovich, S.J. (1711-1787), as seen in this woodcarving with his instruments, was a Jesuit polymath: astronomer, philosopher, and engineer. Most famous for his prescient atomic theory, he also devised the repair of the dome of St. Peter's and succeeded in getting the Copernican system removed from the Index.

firmed today, its popularization as a term referring to rivers or even artificial canals (and to some degree the blame for its misinterpretation) falls on Schiaparelli.

Favorable oppositions of Mars, as Mars and Earth approach each other in their orbits, occur roughly every 26 months (more precisely, two years and 49 days). This drives the calendar for spacecraft missions to Mars, with new missions planned every two years. But even before such missions, these oppositions offered the best opportunities for observing Mars through a telescope.

Furthermore, because Mars has a markedly elliptical orbit, the oppositions that occur near August (which recur about every 17 or 19 years) bring Mars especially close to us, giving it twice the apparent diameter (and thus making it four times as big and bright) in the telescope compared to its appearance during the less favorable oppositions. One such close opposition occurred in 1858, when Secchi made his observations; anoth-er occurred in 1892, when Percival Lowell did his best observing. The most recent was in 2003.

Comparing images of Mars seen through the observatory's *Visuale* telescope in 2003 — recorded with a video chip, not by eye — with the maps made by Lowell, one can see that while there is no evidence of canals, certainly the broader outlines of the dark channels first described by Secchi are very real.

A beautiful illustration of Lowell's map can be seen in a hand-painted globe at the Vatican Observatory made by the Danish artist Ingeborg Bruhn, from 1915 to 1921, based on the drawings in Percival Lowell's books. It is one of five known to exist today. The globe is 20 centimeters in diameter, and it stands some 40 centimeters tall on an elaborate brass stand, on which is inscribed "Mars, after Lowell's Glober [sic], 1894-1914; Thy will be done on earth as it is in heaven; Free Land, Free Trade, Free Men."

("Free land, free trade, free men" was a slogan of the Labor Party, the International Union, and other socialist movements of that era. Early socialist utopians viewed Mars as an ideal place to set up a free society, as exemplified in the novel *Red Star* by Alexander Bogdanov, published in Russia in 1908, the same year as Lowell's final Mars book, *Mars as the Abode of Life*.)

The quotation from the Lord's Prayer, not present on other Bruhn globes, suggests that it was painted specifically for the Vatican. How this globe came into the possession of the

of the first reliable measurements of the grain and bulk densities and porosities of the SNC meteorites, believed to come from Mars (see p. 148). These values, in turn, led to a more precise geophysical modeling of the Martian crust.

At the time that work began, Britt was camera scientist on the Mars Pathfinder mission, which landed both a rover and camera on Mars in 1997. Following up on that mission, he used a sample of the Vatican's SNC meteorite Chassigny (the "C" of the SNC class) to test the microscopic camera that was sent on the (unsuccessful) Mars 2001 lander spacecraft. A similar camera was eventually flown on the successful Spirit and Opportunity 2004 landers.

During his visits to the meteorite collection at Castel Gandolfo in 1996 and 2001, Britt also worked to develop NASA proposals for future spacecraft

Above: Our best look at Mars comes from actually traveling there. The 1997 Mars Pathfinder mission, with its small rover Sojourner, was the first of a series of Mars missions that have visited the Red Planet every two years since its successful landing. In charge of the camera that took this panorama was Specola collaborator and regular visitor Dan Britt, of University of Central Florida. Credit: NASA/University of Arizona.

Vatican Observatory is not known.

The actual study of Mars did not continue at the Vatican Observatory, however, until the 1990s. At that time, a collaboration between this chapter's author and Dr. Daniel Britt (then of the University of Arizona, now at the University of Central Florida) led to some

missions to Mars. He relates his embarrassment when, after a heated (and loud) discussion by telephone with a spacecraft manufacturer in the United States, he suddenly realized that he was in an office, with open windows, directly over the Pope's apartments — while the Pope was in residence!

Deep Impact

The images from Mars Pathfinder arrived in the summer of 1997, during one of the Vatican Observatory summer schools. Eight years later, one of the instructors of that school, Dr. Michael A'Hearn of the University of Maryland, would have his own spacecraft images to show. His mission was to watch a comet nucleus, close-up, as it was impacted by a 350-kilogram copper projectile. The mission was approved in 1999 and successfully hit its target during the summer of 2005.

Where previous Earthbound astronomers had to be content with observing the tails and gaseous comas of comets, this experiment touched, literally, on the ball of ice and dust from which that gas and dust devolves. The results of this mission showed comet nuclei to be dark, very low-density objects that erupt both dust and water ice when impacted, providing the source of their beautiful comet tails. By actually watching the eruption as it occurred, a number of the details of the impact processes could finally be understood.

Saturn and Titan

Deep Impact arrived during the 2005 Vatican Observatory summer school, where Dr. Jonathan Lunine of the University of Arizona served as one of the instructors. The topic of that school was astrobiology and the search for life in the universe. And by that summer, the joint NASA/ESA mission to Saturn, *Cassini*, had succeeded in putting the *Huygens* landing probe onto the surface of its large moon Titan. Dr. Lunine was a member of that team.

Titan is a particularly intriguing place to study for astrobiologists, not because its surface is expected to hold life today, but because it may be a "frozen" example of the kind of organic chemistry that occurred when the Earth itself was very young. Thanks to its great distance from the Sun, more than nine times farther than the Earth, it receives little more than 1 percent of the heat the Earth gets. As a result, the temperature at its surface is -179 C; almost as cold as liquid nitrogen. But still it has a thick nitrogen atmosphere — thicker than Earth's — enshrouded in a mist of methane droplets.

Indeed, the presence of that haze was one of the great mysteries of Titan. Over the age of the Solar System, even the weak sunlight hitting the cloud tops should be enough to destroy all of the atmospheric methane that resupplies the haze. There must be a source of methane within the moon itself to replenish the methane lost in the upper atmosphere.

Lunine and his team found just such a source, in a series of methane "lakes" observed by radar from the Cassini spacecraft. The methane in Titan's atmosphere goes through a cycle that is very reminiscent of the way water on Earth is evaporated, makes clouds, and then rains down upon its surface. We don't know yet how long-lived the lakes are on Titan, nor even how deep they are. But those that our radar can see on its surface are comparable in area to the largest lakes on Earth.

Top: The nucleus of Comet Tempel 1, as seen by the Deep Impact cameras only 67 seconds after it was struck by Deep Impact's impactor spacecraft. Beyond the spokes of light radiating away from the impact site, one can see ridges, scalloped edges, and possible older impact craters on the surface of the comet nucleus. Credit: *NASA/JPL-Caltech/UMD.*

Above: A natural color view of Saturn and five of its icy satellites, made by combining exposures taken by the Cassini spacecraft narrow-angle camera on November 9, 2003, from a distance of 111.4 million kilometers, about three-quarters the distance of the Earth from the Sun, and 235 days before the spacecraft entered Saturn orbit. Credit: NASA/JPL.

Mercury

NASA's MESSENGER (MErcury Surface, Space ENvironment, GEochemistry, and Ranging) space probe promises to revolutionize our ideas about the planet Mercury. It has already twice passed by that body, and it will finally be brought into orbit around the planet in March 2011. Images sent back from the two flybys have shown both clear similarities and intriguing differences between this, the smallest planet, and Earth's rocky Moon.

One member of the MESSENGER team is Dr. Faith Vilas, director of the MMT Observatory in Tucson who was also, along with A'Hearn, an instructor of the 1997 Vatican Observatory summer school, and who served for many years on the board of the Vatican Observatory Foundation. Her particular interest is in determining the mineralogical composition of the Mercurian surface.

The most obvious similarity between Mercury and the Moon is the rough, heavily cratered, airless surface seen in both places. Both are bodies covered with grayish basaltic lava flows that have been peppered heavily with impact craters. But unlike the Moon, Mercury is also covered with huge scarps, cliffs half a kilometer high that run across its surface for hundreds of kilometers. These, it is believed, were formed when Mercury first heated up (and expanded) with the formation of its large iron core — another Mercurian feature not seen in our Moon — and then shrank after the core formation was over and the interior cooled off.

The core itself is believed to be the source of yet another feature

143

Above: *Radar imaging data from the Cassini flyby of Titan provides convincing evidence for large lakes of liquid on its surface. Here the brightness in the image is proportional to how much radar signal is reflected back to the spacecraft, which is generally a sign of how rough the surface is; thus, very smooth surfaces, such as pools of liquid, are very dark. The colors have been added here by tinting regions of low backscatter in blue; radar-brighter regions are shown in tan. The image was then altered, foreshortened to simulate an oblique view of the highest latitude region as seen from a point to its west.* Credit: *NASA/JPL/University of Arizona.*

Above: *This composite view of Titan was built with Cassini images taken on October 9 and October 25, 2006, in an attempt to give a global look at the haze-enshrouded surface of Saturn's largest moon. Looking in infrared wavelengths, which are less affected by the haze of droplets obscuring the surface in visible light, this image was constructed by coloring and combining three images taken at different infrared wavelengths.* Credit: *NASA/JPL/University of Arizona*

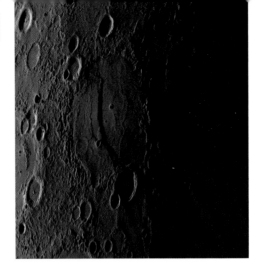

Left: Though looking superficially like the Moon, the surface of Mercury, in fact, shows some significant differences from the lunar surface. One of the most obvious differences is the presence of huge cliffs, kilometers high and hundreds of kilometers long; one of them can be seen cutting across the large circular basin in the middle of this image. This image was taken about 55 minutes before the closest approach of MESSENGER's second flyby of Mercury, on October 6, 2008. Credit: NASA/ Johns Hopkins University Applied Physics Laboratory/Carnegie Institution of Washington.

that makes Mercury different from the Moon: its global magnetic field. Why do some planets have magnetic fields and others do not? The standard answer is that a planet needs a molten metallic core and a strong spin to generate the magnetic field. But Mercury's spin rate is remarkably slow — requiring nearly 60 Earth days to turn around just once. And we're not sure yet if its core is liquid; that is a question that can be approached only by the careful measurement of how Mercury's gravity perturbs MESSENGER once it finally comes to orbit the planet.

Observations done in the 1960s by Fr. George Coyne (who would later become director of the Vatican Observatory and his colleagues at the University of Arizona showed that Mercury's surface altered the polarization of reflected sunlight in a way markedly different from how the surface of the Moon behaved. They suggested this difference indicated that the soil on Mercury had a distinctly different mineralogical composition from that on the Moon. Ten years later, work by Dr. Vilas compared the spectra of the two bodies and proposed that the rocks on the Moon may be richer in iron than those on Mercury.

This trend would be in agreement with some theories for how the chemistry of the planets was determined back when they were formed, 4.6 billion years ago. But when Dr. Vilas gets a closer look at the mineralogy of Mercury's rocks with MESSENGER's instruments only a few hundred kilometers above its surface (rather than with a telescope a hundred million kilometers away, trying to observe Mercury while fighting the glare from the nearby Sun), she should be able to add far more detail to the story.

Above: Shown here are two color images of Mercury taken by the MESSENGER spacecraft on October 6, 2008, from an altitude of 2,500 kilometers above the surface, during its second flyby of the planet. The image on the left was produced by combining images from three filters into red, green, and blue channels, as a general representation of the color seen by the human eye. The right image was created by statistically comparing and contrasting images taken through all 11 narrow-band color filters. This method greatly enhances subtle color differences in the rocks of Mercury's surface, providing insight into the compositional variations present on Mercury and the geologic processes that created those color differences. Credit: NASA/ Johns Hopkins University Applied Physics Laboratory/Arizona State University/Carnegie Institution of Washington.

The Sun

From Mercury, the planet closest to the Sun, we return once again to the star at the center of our Solar System and the closest example of all the stars we study in the sky. The first attempt at understanding the Sun itself as a celestial body, and not merely to understand its motion through the sky, came with the observations of sunspots — a point of heated controversy between Galileo and the Jesuit astronomer Christoph Scheiner. (Ironically, both of their claims for priority were probably superseded by the earlier observations in England of Thomas Herriot.)

In the 19th century, it was again Angelo Secchi who made outstanding contributions to the study of the Sun. He made careful drawings of sunspots and recorded the changes in sunspot activity with time. While making these observations, he discovered solar *spicules*, tubes of gas in the solar photosphere created by the strong magnetic field present there. Perhaps his most outstanding accomplishment was dem-onstrating that the glorious solar *corona* visible from Earth only during a solar eclipse is in fact a part of the Sun and not merely an optical illusion connected with the eclipse. As a sign of the value with which his work is held even today, the five cameras on each of the two NASA STEREO spacecraft ("Solar TErrestrial RElations Observatory"), used to study how material is erupted from the Sun into the corona and on into solar wind, have been named the "Sun-Earth Connection Coronal and Heliospheric Investigation package" (the acronym, of course, spelling SECCHI).

More-recent Vatican astronomers have also played a role in our modern understanding of the Sun. For his doctoral thesis at Georgetown University before joining the Vatican Observatory, Fr. Richard Boyle measured the abundances of a number of difficult-to-determine elements in the Sun's radiating atmosphere. Much of his work was based on observations at the famous McMath solar telescope at Kitt Peak Observatory, near Tucson, Arizona. This work foreshadowed the Vatican Observatory's later establishment of a research group in Tucson.

Today, the most outstanding Earth-based telescope for observing the Sun is the Swedish 1-m Solar Telescope (SST). The SST is operated on the island of La Palma by the Institute for Solar Physics of the Royal Swedish Academy of Sciences in the Spanish Observatorio del Roque de los Muchachos of the Instituto de Astrofísica de Canarias. Participating in the site testing for this telescope was Fr. Juan Casanovas, S.J., who worked at this observatory in the Canary Islands from the mid-1960s to the mid-1970s. Images taken at that telescope, some by former students of Fr. Casanovas (who left the Canaries to join the Vatican Observatory in 1974), have revealed Secchi's spicules in remarkable detail.

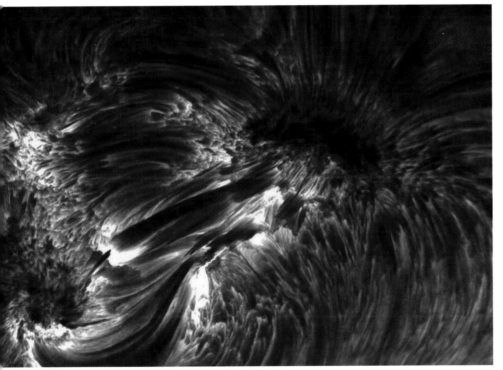

Above: Pictured is perhaps the highest resolution image yet of the solar magnetic flux tubes, called "spicules," first discovered by Fr. Angelo Secchi in the mid-19th century. This image was taken at the Swedish 1-meter Solar Telescope (SST) on the island of La Palma, Spain, on June 16, 2003, as seen through an "H-alpha" filter. The short, dark features at the right are the spicules: jets of gas moving towards us at speeds of ~50,000 km/hour. Visible in the upper left are some small sunspots (where the magnetic field is very strong), and large fibrils or magnetic loops connecting the sunspots. The brighter regions are called "plage regions," also caused by a stronger-than-average magnetic field. The whole area shown here is about five times the size of the Earth. Credit: Institute for Solar Physics of the Royal Swedish Academy of Sciences in the Spanish Observatorio del Roque de los Muchachos of the Instituto de Astrofísica de Canarias.

The Road Forward

Fr. Secchi's book on the Solar System, published 150 years ago, marked a turning point in the study of planets. Up until that time, telescopes were used mostly to pin down the precise location of a

planet so that one could measure its motions, night by night, and work out the mathematics of its orbit. But Secchi, in his introduction, noted that he was also interested in these bodies as places in their own right. He began to work out the sizes and masses of these worlds, and with his telescope he made the first start at mapping out the surface of at least one of them, Mars.

This change in emphasis, from thinking of planets as merely orbiting

point-masses to recognizing them as full and complete worlds in their own right, required two advances. The first was the development of better technology — at first, better telescopes, and ultimately today spacecraft that can actually go and land on these places to study them close-up. But equally revolutionary, it required an advance in our thinking. It required the imagination to recognize that even with limited data, one could begin to see what these worlds must be like.

That revolution in the mind actually took a surprisingly long time to occur. For example, even though Secchi listed in his book the sizes and masses of the moons of Jupiter, he never took the next step of calculating from them just how dense those moons must be, or to ask what materials in space could give the moons a density much lower than any rocky planet. Indeed, it was not until the 1960s that these basic values, available for more than 100 years, were finally used to make the first simple models showing that the moons first discovered by Galileo — Europa, Ganymede, and Callisto — were actu-

Above: The Galilean satellites Europa and Callisto are seen in this true color frame made of narrow angle images taken by the Cassini spacecraft on December 7, 2000, en route past Jupiter on its way to encounter Saturn in 2003. Europa (seen against Jupiter) is 600,000 kilometers above the planet's cloud tops; Callisto (at lower left) is nearly three times that distance at 1.8 million kilometers. Europa is a bit smaller than our Moon; Callisto is 50 percent bigger and three times darker than Europa. Data from the magnetometer carried by the Galileo spacecraft indicate the presence of conducting fluid, most likely salty water, inside both worlds. Credit: *NASA/JPL/University of Arizona.*

Top: This true color picture was made from narrow angle images taken on December 12, 2000, by the Cassini spacecraft as it passed Jupiter en route to its 2003 rendezvous with Saturn. Io and its shadow are shown in transit against the disk of Jupiter. Io is about the size of Earth's Moon; however, it is periodically flexed by Jupiter's gravity as it orbits around Jupiter, producing internal heat that makes it the most volcanically active body in the Solar System, with more than 100 active volcanoes and plumes. The white and reddish colors on its surface are due to the presence of different sulfurous materials; the black areas are silicate rocks. Credit: *NASA/JPL/University of Arizona.*

ally balls of rock and ice, with temperatures in their interiors warm enough to melt that ice and perhaps provide oceans where life could exist. (Io, the fourth Galilean moon and the one closest to Jupiter, has no water ice; instead, it is covered with volcanoes erupting sulfur and frozen sulfur dioxide ices instead.) Among those making these models, including the first published suggestion of life in those oceans, was the present author — during his studies, many years before entering the Jesuits and joining the Vatican Observatory.

The future of planetary sciences will continue to require both spacecraft and imagination. It would be startling if future spacecraft and high-power telescope observations of the planets did not startle us! But only the imagination will be able to weave these new observations into a coherent story that can explain how these planets all got here, and got to be the way they are today. That includes our favorite planet of all, our own home in the cosmos, planet Earth. ●

Above: A Titan Montgolfiere — a hot-air balloon — has been suggested for a future mission to Saturn's moon. Titan is a very diverse world, and a balloon would be able to cover a lot of ground close-up. Most of the topography on Titan is apparently slight, under a kilometer or so. The Titan Montgolfiere would use a radioisotope heat source for both power and to produce the hot air; thus, it would be able to go up and down, even sampling the surface. **Credit:** *Tibor Balint for NASA.*

METEORITES: ALIENS AT THE VATICAN

• Br. GUY CONSOLMAGNO, S.J. •

Introduction

At the Vatican Observatory, you'll find a thousand aliens: meteorites, rocks from outer space that have fallen to the surface of our Earth. In many cases, they'd been seen to fall, making a bright fireball through the air, and collected near craters formed when they'd hit the ground. Others are stray bits of iron, or grayish rock, that don't look like anything from around here. Each bears the name of the place on Earth where it had been found. Some, like Sacramento Mountains, are metallic iron, though rich also in nickel and other metals, etched and polished to show a pattern of interlocking crystals. Some are stone: Agen, like most of these, is made of millimeter-sized balls of rock called *chondrules*, while other stony meteorites, like Nakhla, look like flows of lava from some extraterrestrial volcanism. And a third group, like Fukang, mix iron and stone in roughly equal proportions.

Regardless of their structure, though, their chemical compositions

Right: A globe of Mars (hand-painted by Ingeborn Bruhn c. 1916) shows the original home of the Nakhla meteorite, Mars.

and elemental isotopes differ from any rock on Earth. They really are aliens from outer space.

The core of the collection was put together by a 19th-century French nobleman, the Marquis de Mauroy. Adrien-Charles, Marquis de Mauroy (1848-1927) was a distinguished agronomist and "gentleman-scientist" of the old French nobility, a life member of the Société Française de Minéralogie who served three terms as its vice president. His collection of minerals was famous throughout Europe, and his meteorite collection was said to have been the second-largest private collection in the world. He was a great supporter of schools and scientific institutions; for instance, Czar Nicholas II awarded him the insignia of a Commander of St. Stanislas for his donation of meteorites to the Institute of Mines in Russia.

A great friend of the Church, the Marquis hoped to found a museum of natural history at the Vatican. To that end, he first proposed in 1896 to donate a collection of 1,800 rocks and minerals, and a library of some 400 books and monographs about them, to the Vatican. At that time, however, the observatory (where they were to be housed) had only recently been refounded by Pope Leo XIII and was located in cramped quarters, primarily the Tower of the Winds; the Marquis was thus asked to postpone his donation.

In 1905, a subset of the de Mauroy meteorite collection (about 150 pieces, mostly duplicates and smaller samples) was donated by the Marquis. But his dream of a natural history museum never materialized. When the Vatican Observatory moved from Rome to Castel Gandolfo in 1935, his widow, Marie Caroline Eugénie, donated the remainder of his collection. In 1973, the terrestrial minerals were given in permanent loan to the Geochemical Institute of the University of Vienna; but the meteorites stayed at the Vatican.

Above: The Marquis de Mauroy, who collected the meteorites that became the Vatican collection, and his wife, Marie Caroline Eugénie, who donated the collection to the Vatican at the death of her husband.

In addition to the Marquis's collection, the observatory early on had two other important meteorite donations. An iron meteorite first identified as Angra dos Reis (iron) but now known to be the main mass of Pirapora was donated to Pope Leo XIII and was transferred to the Vatican Observatory in 1917. And in 1912, John Ball, the acting head of the Geological Society of Egypt, kindly sent a 154-gram piece of the newly fallen meteorite Nakhla. Little did he know that by the 1980s this meteorite would become one of the most scientifically exciting falls in our collections, representing one of only a handful of meteorites identified as pieces of the surface of Mars.

Collected over a period of some 200 years, now these "aliens" sit in carefully labeled little plastic bags in drawers in a room in the Pope's summer home: a thousand pieces of outer space. What can they tell us about their origins? What can they tell about the places they've been, and the things they have seen?

Measuring the Meteorites

Starting at about the time of the exploration of the Moon in the 1960s, the chemical study of meteorites has made huge advances. Techniques developed for the study of lunar rocks have been applied to meteorites with great success. Electron microscopes allow one to see the crystals in slices of a meteorite at a scale of better than a millionth of a meter, and the radiation emitted when the electrons hit can tell you, crystal by crystal, what elements are present. In addition, we can now boil off individual atoms from each crystal and look for the particular isotopes produced by the radioactive decay of certain well-studied elements. By knowing how fast the radioactive atoms decay, and counting the number of the daughter isotopes that have accumulated there as the result of those decays, you can calculate how long the crystal has been frozen into its current form and able to collect those atoms. The precision of these measurements is now so good that you can see differences in age of only a million years between rocks that, like most meteorites, are 4.567 billion years old.

But such experiments require expensive and complicated equipment, as well as a trained staff of technicians to keep it all operating properly. When I arrived at the Vatican Observatory in 1993, I realized that we could never duplicate such a lab and stay anywhere within our budget! Besides, those experiments were already well underway in other labs around the world. What I needed to do was to find a range of experiments suitable to the nature of our collection and our limited resources.

Above: A formal portrait of the Martian meteorite that fell near the Egyptian town of Nakhla in 1911. Photo credit: *Alberto Pizzoli.*

The collection had been put together as an amateur's collection: it had mostly small pieces, very few pieces of them "main masses" with material to spare for destructive experiments. But it included a remarkable range of different meteorites and meteorite types; this suggested that I look to do experiments that surveyed characteristics across meteorite types.

On the other hand, I realized that the Marquis had collected most of these samples more than a hundred years ago. Meteorites are known to be filled with tiny flecks of metallic iron; that's one of the chemical traits that distinguishes them from Earth rocks. Earth's atmosphere (and water) attacks this iron, and over time turns it into rust. What sort of measurements could I do that would not be affected by this rust?

Densities

In my earlier career, before I became a Jesuit brother and was assigned to the Vatican Observatory, much of my research had been based on trying to make mathematical models with a computer to describe how small bodies like asteroids and moons evolve over time. There are all sorts of interesting geological processes that can happen even in such small bodies: for example, the insides of icy moons can melt, forming salty oceans between an icy crust and a rocky core — even with the chance that there might be some sort of bacteria (or fish? or dolphins?) swimming around in those oceans! But to make these models, I had needed to know some basic traits about the stuff that moons and asteroids are made from. The characteristics of ice were well known, but the rocky material in these bodies was less well studied. The best guess has always been that whatever rock is out in the Solar System is probably not all that different from the meteorites — which, after all, we know come from that region of space.

So what were the kinds of data I wished I had for my models? I wanted to know how well the rocks collect and conduct heat; how likely they were to flex and change shape under a variety of different stresses. But most basic of all, I really wanted some good numbers for the density of these rocks.

Think about how you used to test the presents under the Christmas tree when you were a kid. You would pick up one wrapped box after another, and judging from its "heft," you could guess which boxes might have chocolate and which ones just had new socks. That heft is, in fact, what we call *density*. Water has a density of one gram per cubic centimeter; iron has a density near eight grams for the same volume. But there are many different kinds of rocks, and their densities can range from two to five grams per cubic centimeter. What's the appropriate value to use when you are making a mathematical model of a moon or asteroid? We know that the meteorites come from the asteroid belt; presum-

ably their density would be what I wanted.

But in fact, it was not at all easy to find a density for many types of meteorites. In part, that was just one of those measurements that no one had gotten around to doing. But digging a little deeper, I began to see why it was so hard to measure.

Above: The Agen H5 meteorite is a typical, ordinary chondrite. It fell in France in 1814. The interior is gray while the outer surface was burned with a black fusion crust by its rapid descent through the Earth's atmosphere. The crust has since turned brown by reactions with Earth's moist atmosphere.

Density is mass divided by volume. You can measure the mass easily enough; just weigh the sample. But volume is trickier, because meteorites are irregular in shape.

Of course, Archimedes had figured out how to measure such a volume some 2,500 years ago. Asked to

test the density of the king's crown (to see if it was really pure gold or just gold covering a less-dense base metal) he pondered the problem while taking a bath; and seeing how the bathwater rose as he was lowered into the tub, he got the bright idea of measuring the volume of the water spilled out of a full bucket when the crown was immersed in it. (He was so excited at this idea, the story goes, that he jumped from his bath and ran naked down the streets of Syracuse, shouting "Eureka!") A modern variation of this idea — mathematically it comes to the same thing — is to weigh the rock first in air, dangling from a string, and then see how the weight changes when the rock is dangled into water.

But I could see, there was trouble dipping meteorites into water. I knew that my meteorites were liable to rust just sitting in air. To dunk them in water risked doing permanent damage to their chemistry — not to mention the risk of contaminants from the water getting into the rock, which might invalidate any future chemical measurements. (Modern probes are so sensitive that you need to remove your rings before handling meteorites, for fear of stray elements from the metal jewelry contaminating the samples.)

To prevent such contamination, perhaps you could wrap the meteorites in plastic. But I tried that out on a pile of sugar cubes (being cubes, I could measure their volume directly), and no matter how tightly I tried wrapping the plastic, the cubes tended to come out soggy. I read papers where scientists in Japan had actually just carved some of their meteorites into perfect cubes; but I didn't want to carve up my samples, and besides, how could I tell if that cutting might not change the structure of the meteorites, introducing internal cracks that might change their density?

But then I had my own "eureka" moment over a cup of cappuccino. Every morning at ten o'clock, all work stops at the observatory, and we gather in the kitchen of the Jesuit residence to take a coffee and chat with one another about our work (and the local football team). It's one of the perks of living in Italy. But while I was pouring my sugar into my cappuccino, it suddenly occurred to me that a powder like sugar behaved much like a fluid. If I poured it over my meteorites, it would fit evenly around all the irregular corners and crannies, but it wouldn't actually react with or contaminate the rock.

Take a plastic measuring cup, measure its volume, then fill it with your powder and weigh it. The weight, divided by the volume, tells you the density of the powder. Now, insert the meteorite in the cup, fill the rest of it with the powder, and weigh it again. The difference between the two weights, with a little bit of algebra, can eventually lead to the density of the rock compared to that of the powder. I tried it out, and it seemed to work!

Later, visiting our observatory's Tucson offices, I described my method to my friend and colleague, Dan Britt, who was working then with the Mars Pathfinder mission at the University of Arizona's Lunar and Planetary Lab.

Dan was interested in meteorites because he wanted to compare their densities to asteroids. From missions to Mars, we had gotten density values for the little Mars moons, Phobos and Deimos, which everyone agreed are probably captured asteroids. And

the Galileo spacecraft had obtained a density for asteroid Ida while passing through the asteroid belt on its way to Jupiter. (Galileo's images gave us a measure of Ida's volume, while a series of pictures showed the motion of its little moon Dactyl, which could be used to calculate its mass.) In the next few years, a number of other missions to asteroids were being planned. But what good is an asteroid density if you don't have any meteorite densities to

compare them against?

We compared the different ways of measuring meteorite density. "The trouble with using water isn't just the contamination," he pointed out. "Rocks, like meteorites, can be riddled with cracks and other pore spaces. The water gets into some of those spaces, but you never know just how much of the pore spaces get filled. So I was talking to a geologist working on measuring the density of core samples.

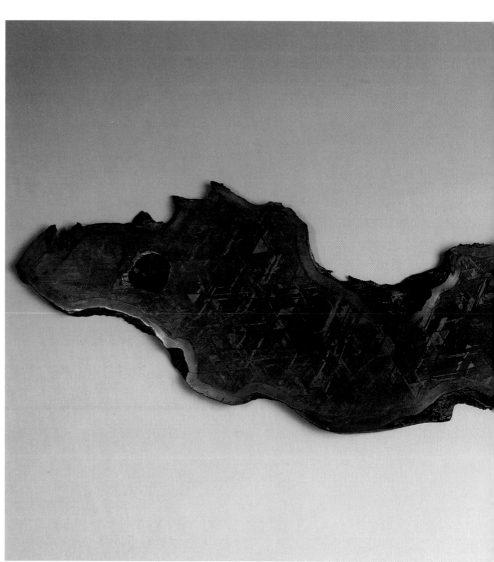

Above: Iron meteorites, like this large slice of San Francisco Mountains, show a complex crystal pattern when etched with acid. The crystals indicate the high nickel content in the metal. Photo credit: *Alberto Pizzoli.*

They use a device called a pycnometer. Instead of water, it uses helium gas. Helium is completely inert, so there's no contamination."

He explained how it worked. The rock is sealed in a chamber of known volume, flushed with helium at room pressure. A second chamber of known volume with helium at, say, two atmospheres pressure, is attached to the first chamber. Open the valve between them and measure the final pressure. The bigger the rock, the more space it takes up in its chamber, the less room there is for helium, and so the higher the final pressure.

"But the best part," Dan told me, "is that the helium gets into *all* the cracks. It tells you the volume of just the rocky part of the rock."

But I realized that my powder method would measure the volume of the rock including any pore spaces. "Isn't that the volume of the meteorite we really want, to compare with asteroids?" I wondered.

"Actually, you want both," he replied. "The difference between the two volumes is the volume of the pore space: this way, you can measure the porosity of the meteorites."

Sand on a beach is 50 percent empty space. Typical sandstones can be as much as 30 percent pore space. "I've searched the literature for porosity measurements of meteorites," he told me, "and they're darn hard to find. One of the most common classes of meteorites, LL chondrites, has had only two porosities published. One was 3 percent; the other was 30 percent. Which are we supposed to believe?"

Once I had explained my powder method, Dan immediately went about improving it. Instead of plastic cups and sugar, he obtained flat-topped beakers and 40-micrometer-diameter glass beads. Unlike other powders (like sugar) the beads were round and poured around the rocks much more smoothly.

In April 1996, I invited Dan and his family to visit us in Castel Gandolfo; he brought the pycnometer, and together we set up the lab. For the next three months, I measured nearly a hundred different meteorite samples from the Vatican collection. The first time I presented the results, at the annual meeting of the Meteoritical Society in Berlin, a grand old man of the field took me aside. "Why are you doing density measurements?" he asked me. "Nobody does that!" That, I thought, was precisely the idea.

But as we gathered more data and began to see patterns in the porosity, eventually people began to get interested. Our first results were published in the journal *Meteoritics and Planetary Sciences*. Soon our technique, and our measurements, became a standard source for the community. Ten years later, we were invited to review our work in a lengthy article for the prestigious journal *Chemie der Erde*. It's quite ironic that we would publish in a journal whose title translates as "chemistry of the Earth," since our measurements were about the physical nature, not the chemistry, of our samples — and those samples were definitely not part of the Earth!

This great interest was spurred in no small part by fortuitous good timing. Not only were spacecraft measurements providing a handful of asteroid densities, but just as our work was getting published, improved telescope techniques on Earth had led to the discovery of dozens of asteroids with small moons. Once you can see the asteroid pulling the moon around itself, you can calculate its mass — the hardest number to get for an asteroid density. And it turned out, these bulk density measurements indicated that compared to our meteorites, these asteroids were 20 percent to 50 percent empty space: porous on a scale bigger than the porosity within the meteorites themselves. Asteroids are not solid rocks orbiting the Sun; they are, at the least, heavily fractured bodies with deep cracks running through them. And some of them are probably piles of rubble, as porous as a bucket of beach sand.

One clear trend in our data is that bodies larger than about 10^{20} kilograms in mass, or about 500 kilometers in diameter, have no porosity at all. They must have enough gravity to pull themselves into a spherical shape. This could be the basis for defining the boundary between small Solar System bodies like asteroids and comets, and the newly defined class of Pluto-like bodies called dwarf planets.

Digging deeper into the data, recently we have begun to see an interesting trend. The asteroids from the inner part of the asteroid belt tend to be about 20 percent less dense than the "ordinary chondrite" meteorites, which they resemble; and those meteorites are full of microcracks that add another 10 percent of porosity. But asteroids farther away from the Sun are much darker, like "carbonaceous chondrite" meteorites. They are also much more porous — as much as 50 percent empty space — and the meteorites themselves have another 25 percent porosity. As one goes to the colder parts of the Solar System, solid material gets to be quite fluffy indeed. And this is confirmed by a handful of measurements indicating that the nuclei of icy comets may be as much as 80 percent empty space!

Not only does such a structure for an asteroid have a profound sig-

in an eccentric orbit could someday hit the Earth. We certainly see small bits of asteroids hit Earth today: the meteorites themselves, and more commonly meteors or "shooting stars," as Fr. Kikwaya describes in his chapter. Smaller ones are more common, but bigger ones do hit. But if we knew that such an asteroid was heading our way, how could we deflect it? Hollywood's answer is to send Bruce Willis up with a big bomb. But a rubble pile is

Above: Stony iron meteorites, like this slice of Fukang, combine basaltic minerals with iron. The translucent green crystals visible here are olivine (known as peridot when in gem form), while the metal has a composition similar to the iron meteorites. That both the dense metal and the less dense olivine crystallized together without separating indicates that these meteorites were formed in a region of very low gravity.

nificance for our understanding of how the solid material that eventually made up planets like Earth was processed in the early Solar System, but it also has very practical implications.

As a number of Hollywood movies have tried to dramatize, there is a real (if small) threat that an asteroid

already broken into pieces; a bomb will just absorb the blast, squeezing and rearranging the rubble without dispersing anything. A more reasonable strategy might be to put a small device on the asteroid surface that gathers up the loose rock and spits it out in one direction, slowly deflecting the rest of the asteroid the other way and thus moving it out of its collision path.

Pennies from Heaven

But asteroids are more than just a threat. As we noted, there are many lines of evidence suggesting that ordinary chondrite meteorites can be derived from a particular type of asteroid, the "S-class," found mostly in the inner asteroid belt. We also know that many such asteroids are in orbits that pass near the Earth. Through 2008, about 5,500 such asteroids have already been discovered; 750 of them are larger than one kilometer in diameter. Many of them may at one time or another pass as close to the Earth as Earth's Moon orbits now, a distance that we know we can traverse with manned spacecraft. These are the ones that Hollywood views as threats.

But consider an asteroid of 10 kilometer radius. The typical S-class asteroid has a density of about 2,500 kilograms per cubic meter; so the total mass of one such asteroid is roughly 10^{16} kilograms. If its composition is the same as an ordinary chondrite, it will be about 10 percent metallic iron and other siderophile (metallic) elements.

Ten to the fifteenth kilograms of iron — 1 trillion metric tons — is a thousand times greater than the entire annual output of iron ore everywhere on Earth. The other metallic elements present in such an asteroid, such as gold or platinum, would likewise overwhelm domestic demand for such metals.

A mining expedition that goes out to collect valuable minerals from the asteroids needs to know what sort of surface to expect. We know now that it won't be solid rock. But will the pieces of rubble be grains of dust, or blocks the size of houses? We're still puzzling over that one.

This calculation does show, however, that perhaps the best way to remove the threat of an incoming asteroid is to remove the asteroid itself, bit by bit.

And perhaps someday the Vatican's meteorite collection will be supplemented not only by new samples seen to fall from Earth, but pieces that we've actually gone out and fetched from space itself. Someday, the aliens in our collection may not be immigrants, but souvenirs. ●

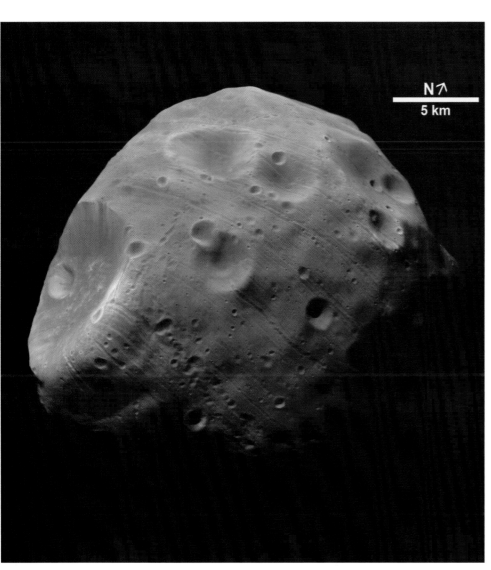

Above: Phobos, the larger moon of Mars, is probably a captured asteroid. You can see evidence for a significant amount of cracks and voids; notice the number of parallel grooves. When you compare the density of small Solar System bodies like this one with that of the meteorites we believe come from those asteroids, it becomes evident that most have significant internal voids, and may be rubble piles. Image taken by the High Resolution Stereo Camera on board ESA's Mars Express spacecraft on August 22, 2004. Credit: ESA.

Br. Guy Consolmagno, S.J. (USA), is the curator of the Vatican Meteorite Collection at the Vatican Observatory in Castel Gandolfo. An expanded discussion of this topic can be found in his book Brother Astronomer, *published by McGraw-Hill in 2000.*

THE PHYSICAL NATURE OF METEORS

• Fr. Jean-Baptiste KIKWAYA, S.J. •

Above: Comet Whipple-Fedtke-Tevadze (C/1942 X1), photographed at the Vatican Observatory on March 1, 1943. Compare its tail to that seen one month earlier, on page 33. Many meteors come from comet tails.
Opposite below: The great show of shooting stars of November 27, 1872. Woodcut from C. Flammarion, Astronomie Populaire, Paris: Marpon et Flammarion, 1880; p. 257.

The Bombardment of the Planets

Studies of the proportional abundances of decaying radioactive elements in rocks from the Earth, the Moon, and the meteorites that fall to Earth from the asteroid belt all tell us that these elements were fixed into the minerals of these rocks some 4.6 billion years ago. This is generally accepted as the age of the Solar System, when the Sun and planets were formed out of a cloud of gas and dust.

But some 400,000 to 800,000 years after the Solar System formed, or 4.2

METEORS

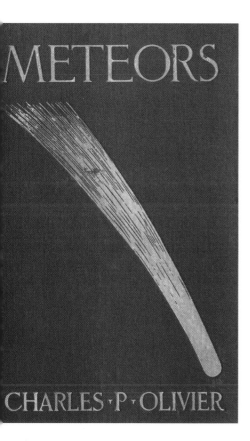

CHARLES · P · OLIVIER

occurred as Jupiter's gravity swept material left over from the formation of the planets, especially stuff from the asteroid belt, into orbits that would pass by the inner planets on their way to being absorbed by the Sun. According to this theory, 22,000 or more impact craters with diameters larger than 20 km, about 40 impact craters with diameters about 1,000 km, and several impact basins with diameters about 5,000 km would have formed on the Earth, leaving its face as pockmarked as the Moon. Evidence for such cratering has also been seen on the surfaces of the other inner planets (Mercury, Venus, and Mars).

Phenomenon like the Late Heavy Bombardment, able to cause such environmental damage, may never occur again. Still, bodies of different size, mass, composition, and origin continue

a "size considerably smaller than an asteroid and considerably larger than an atom" according to the definition from International Astronomical Union. Their size varies from 10 micrometers to 50 meters. When crossing the Earth's atmosphere, the contact with the air molecules turns them to brightly visible events called *meteors*. Depending on their size and composition, meteors can totally ablate in the atmosphere, or survive partly the air friction and end on the ground as *meteorites*.)

The evidence of the occurrence of meteors in ancient human history is available in Chinese and Japanese records, and in E. Biot's 1846 catalog of *étoiles filantes*. The oldest possible meteor activity, suggested by the sentence "many stars flew, crossing each other," goes back as far as the 19th century B.C. Another occurrence of mete-

billion years to 3.8 billion years ago, the Moon apparently experienced an immense shower of meteoritic debris raining upon its surface, judging from the dating of lunar samples, which indicates that most of the impact melt rocks from there formed within this slice of time. This has been called the Late Heavy Bombardment, or Lunar Cataclysm.

The oldest known rocks on the Earth also place the solidification of the terrestrial crust at around 3.8 billion years ago. Was the Earth's crust molten all the time before then? While some theories suggest that the Earth's crust could have formed molten, and stayed molten for up to the next 0.7 billion years later, others suggest the Earth-sized mass should have solidified even as quickly as 4.5 billion years ago. (And even with a slowly-cooling Earth, there ought to be some rocks older than 3.8 billion years on Earth.) The coincidence of ages likewise suggests the existence of a Late Heavy Bombardment at about this time.

This bombardment apparently

to fall onto the Earth. Meteorites from this rain can be seen in our laboratories and museums. But the most visible traces of this shower of material are the meteors: the "shooting stars."

(A little clarification on terminology: *meteoroids* are these objects moving in the interplanetary space and having

ors might possibly date from the flight of the children of Israel from Egypt around 1440 B.C.! C. P. Olivier, in his 1925 book *Meteors*, suggested that the metaphor "a pillar of cloud by day ... a pillar of fire by night" (Exodus 13:21) describes the daylight dust clouds and nighttime glowing trains left by extremely large meteors. He also proposed that another verse in the book of Joshua (10:11) could be interpreted in connection with meteor activity: "The LORD threw down huge stones from heaven." While these stones are later called hailstones, Olivier suggested that they could have been meteorites. From then to now, meteors of different size, mass, structure and composition still occur.

Our scientific understanding of meteors in physics and astronomy began in the 19th century. Many different meteor showers have been identified, and the link between these streams and their parent bodies was first made by Schiaparelli in the 1860s. More recently, new instruments (optical and radar) have been built to provide an ever more accurate knowledge of meteoroid dynamical, physical, and chemical properties.

The Phenomenon of Meteors

When a meteoroid penetrates the Earth's atmosphere, depending on its mass and velocity it can create one of the four phenomena: *meteor*, *fireball* (or *bolide*), *explosive impact*, or *meteoric dust particle*.

a) Meteor

During its flight in the Earth's atmosphere, a meteoroid of a size between 0.01 mm and 20 cm is heated by friction with air molecules. The surface temperature rises about 2200 K at a height of around 100 km (the region of Earth's atmosphere known as the *mesosphere*). The meteoroid starts to ablate from the surface, filling the surrounding atmosphere with excited atoms. The gradual de-excitation of these atoms produces light radiated in the visible bandwidth, creating the phenomenon called a *meteor*. Through the process of ablation, the meteoroid will lose all its mass and, after travelling a distance of several kilometers, will terminate its flight.

The dust that forms such meteors must come from someplace. But where? Small dust particles dating from the origin of the Solar System, 4.6 billion years ago, would have long since been swept out of the Solar

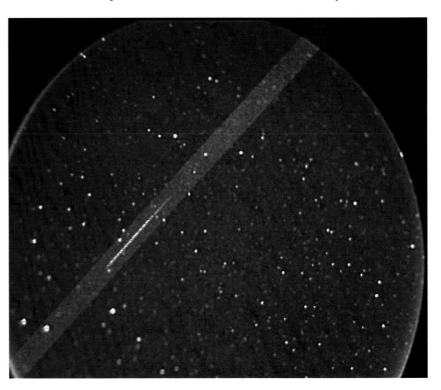

Above: Meteor captured with Deep GenII camera at Silo (London, Ontario) on May 27, 2004 at 08:19:27 UTC.
Opposite: Perseid captured with GenIII camera at Elginfield (London, Ontario) on August 13, 2007, at 08:12:29.840 UTC (meteor appears as a series of regular dots, which help to compute the velocity and the deceleration).

System just by the pressure of sunlight. So we assume the dust originates from somewhat larger bodies: comets and asteroids.

But which meteors come from comets, and which from asteroids? F. Whipple in 1954 introduced an empirical criterion to associate meteor streams to their parent bodies, comets or asteroids, and many authors started to apply his criterion to individual sporadic meteor orbits. Others criticized this attempt, arguing that Whipple's criterion is an empirical boundary line between cometary and asteroidal orbits that could hardly be extrapolated down to distances crossing the Earth's orbit.

Those meteoroids that last only several tenths of seconds in the Earth's atmosphere were thought to have the same velocity (around 60 km/s) and the same origin (presumably from comets). However, close inspection of these meteors has shown that their meteoroids have different compositions, densities, and structures, with the result that they have different behaviors during their flight in the Earth's atmosphere.

Knowing that individual groups of sporadic meteors could have not only different orbits but also different compositions, Z. Ceplecha, in 1967, suggested using another empirical expression more suitable to distinguish between cometary and asteroidal origins, which included a parameter representing a measure of the heat conductivity, density, and specific heat of the meteoric particle. Studying photographs of 2,000 meteors with orbits negligible number arriving in highly inclined orbits, and they come from orbits near the Earth. Group B meteors come from between Earth and Mars, and are somewhat tilted from the ecliptic. Group C includes an ecliptical concentration but also has a considerable number of very highly inclined orbits. Some of these meteors also appear to have originated from much farther away from the Earth, with orbits more like comet orbits.

of such meteoroids had been captured. This led to a further subclassification of the Group C meteors.

Finally, working on 3,624 meteors, combining photographic and TV images made at several stations — *double* and *multistation* meteors — Ceplecha, in 1988, found another class of meteors he called Group D.

Each of the four groups (A, B, C, D) has a different density. Category A meteoroid densities range from 1400 to

and the other data needed to compute this parameter, Ceplecha noticed that small meteoroids show differences that can be used to classify them into different groups.

Group A meteors have orbits concentrated in the same plane as the planets' orbits (the *ecliptic*), with a

Using a television (TV) camera system, R. L. Hawkes collected images of 84 double-station meteors during the autumn of 1976. Being more sensitive than previous systems, the TV system allowed the detection of meteors with brightness fainter than +3 magnitude. It was the first time ever that images

2700 kg/m³ with a characteristic ablation coefficient, which is an expression combining drag (how much the oncoming flow through the air leads to the deceleration of the body), heat-transfer (how much of the kinetic energy of the oncoming stream of molecules is used to burn away a certain mass of the body), and the energy that must be delivered to a certain mass of the meteoroid in order to heat it from its initial temperature to its evaporation or melting temperature. Category B densities range from 650 to 1700 kg/m³ and have a somewhat higher ablation coefficient. The density of category C meteoroids covers the range from 550 to 920 kg/m³ with yet a higher ablation coefficient. The last class of small meteoroids, D, has densities between 180 and 380 kg/m³ and the highest ablation coefficient. The fact that each group has a different value of density and ablation coefficient shows that small meteoroids are different and belong to different groups.

The last aspect we can mention about small meteoroids is their structure. Small meteoroids can undergo fragmentation, which occurs mostly under the effect of the temperature on the surface of the body. Looking at the density of very small meteoroids, my own work (with colleagues at the University of Western Ontario) has tried to reconcile two different approaches of the modelers: those whose work on density is mainly based on fitting the meteoroid light curve taking into account meteoroid fragmentation, and others who considered the meteoroid as a single body that would not undergo fragmentation but could decelerate. We have been working on six double station meteors captured with a new generation of TV cameras (GenII), whose field of view is six degrees, capable of showing their deceleration. Working on a model of ablation that takes into account all processes of ablation of the meteoroid, we have fitted both light curves and the deceleration of the meteoroids. We found that three meteors had high density, close to iron, and three meteors had low density. We suggest that the flux of ironic meteoroids might be underestimated (Ceplecha estimated them to 1 percent). These results are close to those of recent observers who analyzed 97 meteoroid spectra and found that 12 percent are iron.

b) Fireballs

A meteoroid larger than 20 cm traveling at the velocity of 15 km/s doesn't ablate entirely, because its velocity drops along its trajectory during its flight in the Earth's atmosphere, reaching a point where the transfer of energy to the body cannot keep its surface at 2200 K or more. The remnant will start to cool down; the surface will solidify and form a crust typical for meteorites. But while it is hot enough to glow, the large meteoroid shines with a magnitude of minus eight or brighter (comparable at least to the brightness of the crescent Moon). This phenomenon is called a *fireball*. After it ceases to emit light, it continues in a *dark-flight*, which lasts longer than the luminous one, and its terminal velocity approaches the theoretical free-fall velocity of a falling body in a resisting medium.

Studying two different fireballs named Sumava and Benesov with similar initial velocities (27 km/s and 21 km/s), J. Borovicka noticed that their behavior in the atmosphere was very

Above: A fireball imaged in October 2008 at the Okie-Tex Star Party. Photo credit: *David Wang.*
Opposite: Microphotograph of an interplanetary dust particle. Photo credit: *Scott Messenger, NASA Johnson Space Center.*

different. Sumava reached a maximum brightness magnitude -21.5 (about 1 percent as bright as the Sun!) at a height of 67 km, while at that height Benesov was still in the process of its initial increase of luminosity. At 58 km, Sumava had completely ablated, but Benesov continued to radiate below 20 km. He concluded that these observed differences were due to the different compositions, densities, and structures of each of the meteoroids.

To classify fireballs into different groups, Ceplecha used another experimental parameter based on the initial velocity when the meteoroid hits the Earth's atmosphere, the air density at the terminal point of the fireball luminous trajectory, and the initial meteoroid mass in grams. He found that fireballs could be classified into three different groups (plus various subsets of those groups) and deduced the average bulk densities for these different meteoroid groups.

In October 2008, for the first time the arrival of a fireball was actually predicted. A survey searching for small Near Earth Asteroids discovered a meteoroid, designated 2008 TC3, and it was immediately recognized that its calculated orbit would soon intersect the surface of the Earth. One day later (October 7), as predicted, the flash of its impact was observed over Sudan by weather satellites.

c) Explosive impact

The third phenomenon created by a meteoroid entering the Earth's atmosphere is an explosive impact, when the object hits the Earth's surface. Such a meteoroid must have a size of several meters; it must be sufficiently strong; and it must have a considerable mass. It hits the Earth's surface at a hypersonic velocity (several kilometers per second, faster than the speed of sound in the rock it hits), emitting light and creating an explosive crater.

Perhaps the most famous impact crater is the 1.2 kilometer-wide Meteor Crater in northern Arizona, which was formed about 50,000 years ago by the impact of a 300,000-tonne iron meteoroid. But craters continue to be formed even today. On September 15, 2007, at 4:45 p.m., an ordinary chondrite meteorite believed to be at least 1 meter across struck southern Peru. Upon its collision with the Earth, it formed a circular crater of approximately 30 meters wide and 6 meters deep. Eyewitnesses recalled seeing a large ball fall from the heavens with the sound of a jetliner. Some testimonies reported water boiling in the ground 10 minutes after the impact.

d) Dust particle

At the other extreme, a meteoroid with a size less than several hundredths of a millimeter does not produce light, and therefore it is not observed as a luminous phenomenon. While it is still high in the atmosphere, it slows down to a few kilometers per second and, consequently, doesn't reach the evaporation regime. It is called a *meteoric dust* particle, or *micrometeorite*, and it makes its way to the ground through the Earth's atmosphere, sedimenting slowly and remaining unchanged.

It is believed that organic interplanetary materials may be brought to the surface of the Earth by meteoric dust particles. It is estimated that 30,000 tons of interplanetary dust accretes onto the Earth every year. Unlike the first category of meteoroids described above (larger particles of about 400 micrometers, heated during their atmospheric deceleration and vaporized as meteors), meteoroids of 10 micrometers or less don't ablate, and they might contribute a bulk mass that is not heated above 600 degrees C (the pyrolysis temperature of extraterrestrial organic matter). Collected from the Earth's stratosphere, such particles have been shown to contain a high abundance of carbon, including percent-levels of both carbonyl ($C=O$) and aliphatic hydrocarbons ($C-H_3$ and $C-H_2$).

Conclusion

The fall of meteoroids onto the Earth didn't stop with the Late Heavy Bombardment, but it is still happening even up to now. Estimates suggest that 5 million tons of sporadic meteors strike the entire Earth's surface each year, while the accretion of interplanetary dust onto the Earth may reach a value of 30,000 tons per year. These meteoroids have different sizes (from less than 100 micrometers to several meters), different compositions and different structures. The study of their physical characteristics reveals to us the secret of the early Solar System; and some scientists think that the accretion of interplanetary dust supplied the Earth with interplanetary organic materials from which life arose. ●

161

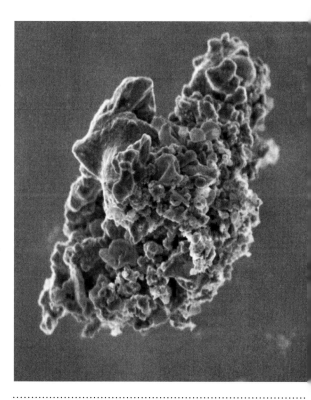

Fr. Jean-Baptiste Kikwaya Eluo, S.J. (Democratic Republic of the Congo), a specialist in celestial mechanics, is a member of the Vatican Observatory now studying at the University of Western Ontario. His present work is determining the physical properties of meteoroids from the study of their dynamics.

EXTRASOLAR PLANETS

• Fr. Giuseppe KOCH, s.j. •

Do you remember *The Little Prince* by Antoine de Saint-Exupéry? Let's imagine a trip like his to *Proxima Centauri,* the star closest to our Sun. Looking back into the night sky in the direction from which we came, the Sun would appear to be just another tiny star. No one would ever expect that it could have planets around it, and that on one of those planets you would find intelligent life, and that it is home to us.

Likewise, we don't know if *Proxima* has a similar array of planets. The distances are too large. Until just a few years ago this seemed to present an insurmountable problem to the study of planets outside our Solar System (*extrasolar planets* or *exoplanets*). But astronomers are very inventive, and won't give up. With the development of various means of research, and the continual advance in precision of our instruments, the exoplanet hunters have finally realized their dream. In 13 years, from the first discoveries in 1995 through the end of 2008, some 333 extrasolar planets have been found around some 260 different stars.

On October 6, 1995, in Florence, not far from the Arcetri hillside home of Galileo (he of the "Medici Planets"), the astronomers Michel Mayor and Didier Queloz of the Observatory of Geneva announced to the Ninth Cambridge Workshop that they had discovered the presence of the planet *51Peg b,* the first known planet orbiting a star outside our Solar System. It orbits around a star very similar to the Sun, *51 Pegasi* (51 Peg), lying 48 light-years away. The discovery was made thanks to an advanced-technology spectrograph attached to the 1.93-meter telescope at the Observatory of Haute Provence.

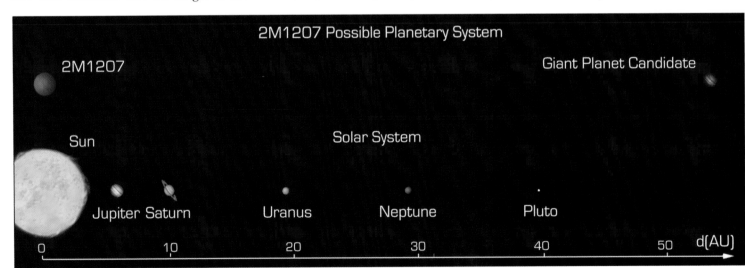

Above: This figure, provided by the European Southern Observatory (ESO), compares our solar system with the system inferred to orbit about the Brown Dwarf star 2M1207. The sizes of the objects are drawn to the same scale, but the distances have been strongly compressed. From the infrared colors and the spectral data available, they conclude that this star is orbited by a 5 jupiter-mass planet, about 55 AU from 2M1207 (55 times more distant than the Earth is from the Sun). However, unlike a planet at such a large distance from our Sun, the surface temperature of this object appears to be about 10 times hotter than Jupiter, about 1000°C. The explanation is that this very young planet is still contracting and releasing energy from its interior. (Jupiter itself, though much smaller and much older, is still releasing gravitational energy from its interior in this way, though not nearly as much.)

Following the announcement in Florence, other groups of researchers turned their instruments towards 51 Peg and soon confirmed this discovery. The first of these, after only a few days, were a pair of American astronomers, Geoffrey Marcy and Paul Butler, who observed at the Lick Observatory of the University of California at Santa Cruz, and who, like their Swiss colleagues, had been working for years to find extrasolar planets. These two teams and their colleagues have dominated the scene of exoplanets; in the decade since the announcement of *51Peg b* they have been responsible for the great majority of discoveries. The official IAU list of exoplanets as of February 2005 counted 127; only 16 were found between 1995 and 1999, and for the first ten years these discoveries occurred at a rate of roughly one per month. But 150 more have been found in the last three years. In the third week of June, 2008, researchers announced the discovery of nine new exoplanets, of which five were so-called "super-Earths," planets that may be terrestrial in nature with masses only a few times larger than Earth's. Seeing how the rate of discovery is accelerating, it won't be long before we find a true "twin" of our planet.

What Can We Say about 51 Peg and Its Companion 51 Peg b?

51 Peg is a 5.5th-magnitude star, of a spectral type quite similar to the Sun, and therefore its color is yellow-white, and

it has roughly the same mass and luminosity as the Sun. It is located 48 light-years from us (our Sun is 8 light-minutes away from us, about 150 million kilometers). It is located in the sky near the western edge of the Great Square, which makes up the wing of the constellation Pegasus. According to Greek legend, Pegasus was the winged horse that was formed from the blood of Medusa when she was beheaded by Perseus. The constellation is easy to find in the northern sky, attached to a large square of stars situated on the opposite side of Polaris (the North Star) from the Ursa Major, the Big Dipper. It dominates the eastern sky in autumn.

51 Peg is a typical G-type star; but its companion, the planet 51 Peg b, was full of surprises for the astronomers. Its estimated mass is about half the mass of Jupiter (this is at minimum; because of the way the planet was discovered, a larger mass cannot be ruled out, but it is probably half the mass of Jupiter). That makes it about 150 times more massive than the Earth. And it lies in a circular

orbit around 51 Peg. But the period of that orbit is only 4.23 Earth days! That means it must be only about 7 million kilometers from its mother star — eight times closer to it than Mercury is to our Sun. It was completely unexpected that a "giant" planet, comparable to Jupiter, would be found at a radius only 5 percent of the Sun-Earth distance, about a hundred times closer than Jupiter is to the Sun. The nearness of 51 Peg b to its mother star implies that it must have a surface temperature of around 1500 degrees. There is no possibility of life on a planet like that. We now call 51 Peg b and planets like it "Hot Jupiters," or, more generally, Extrasolar Giant Planets (EGPs).

According to all the models for planetary system formation current at the time it was discovered, nobody had ever thought that one could find such a massive planet so close to its "sun." Ever since 51 Peg b came on the scene, theorists have been frantically busy revising their theories on the origin and evolution of stars and their planetary systems.

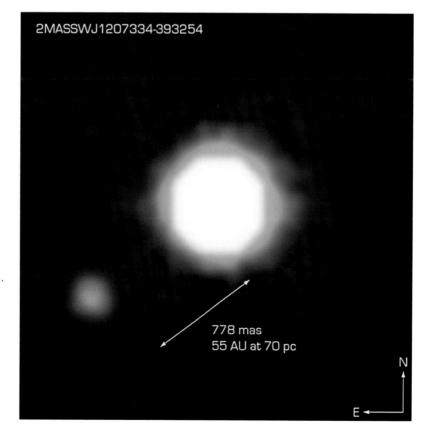

Right: This composite image of the brown dwarf object 2M1207 (center) and the fainter object seen near it, at an angular distance of 778 milliarcsec, was released by the European Southern Observatory (ESO) in 2006. It may represent the first image of an exoplanet, at least in infrared wavelengths. The photo is based on three near-infrared exposures (in the H, K and L' wavebands) with the NACO adaptive-optics facility at the 8.2-m VLT Yepun telescope at the ESO Paranal Observatory.

What Do We Know about the Formation of Stars and Their Planetary Disks?

In the standard model of the formation and evolution of the universe according to the Big Bang paradigm, the formation of stars came in a sequence of phases as a part of the contraction due to gravity of extensive clouds of gas, present as atoms and molecules. In the case of "first generation" stars, this means exclusively hydrogen and helium. For stars of second and successive generations, the clouds forming the star also contained micrograins of dust, formed of several million atoms, heavier elements left over as the residue of the previous generation of stars.

Once the protostar is formed, the cloud in the next phase is present as a somewhat flattened disk of rotating material: the so-called *protostellar nebula* (*solar nebula* refers to our planetary system).

It is generally accepted that such a circumstellar disk plays a fundamental role in the formation of stars and that planets are formed from within it. Before the discovery of 51 Peg b, the theories of such disks were based only on the unique data point of our own Solar System.

The material at the center of the disk begins to condense into a star while a small part of the nebula gives rise to the planets, in the coldest and most distant parts of the system. In our Solar System, the rocky planets — Mercury, Venus, Earth, and Mars — and the asteroid belt are situated relatively close to the central star; the gas giants are much farther away from the Sun. Jupiter and Saturn are, respectively, five and ten times as far from the Sun as the Earth-Sun distance (an *astronomical unit*, or AU). In the case of our planetary system, analyzing the chemical composition of the Sun and the planets and comparing from one to the next the percentage of trace elements heavier than hydrogen and helium, you can ascertain that each "metallic" element is present with the same relative abundance. This is one clear indication that the star and the planets came out of the same nebula, the same "quarry" of original material.

Models of formation and evolution originally proposed for the Solar System ought to be able to provide an

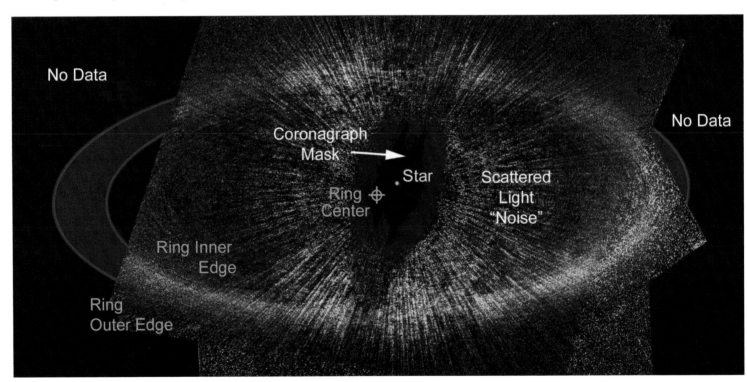

Above: This image, taken by NASA's Hubble Space Telescope, shows a dust ring around the nearby, bright young star Fomalhaut. The center of the ring is about 15 AU away from the star. The dot near the ring's center marks the star's location. Astronomers predicted that a planet, at that time still unseen, moving in an elliptical orbit was reshaping the ring, orbiting some 50 to 70 astronomical units from the star. Credit: NASA, ESA, P. Kalas and J. Graham (University of California, Berkeley), and M. Clampin (NASA's Goddard Space Flight Center).

explanation for the observed properties of the new solar systems discovered around stars other than the Sun. The study of the formation and evolution of these systems range from computer simulations to the observations of several tens of protoplanetary disks, which can be observed in our telescopes in various phases of their evolution. It is just as if you were to take examples of various children of various ages from different parts of

The discovery of the first exoplanet was rapidly followed by an overwhelming series of successive discoveries. It was not just an increase in the number of individual planets, but the rise of many different lines of research that followed, one after another.

In 1999, the first *system* of extrasolar planets was found to orbit about

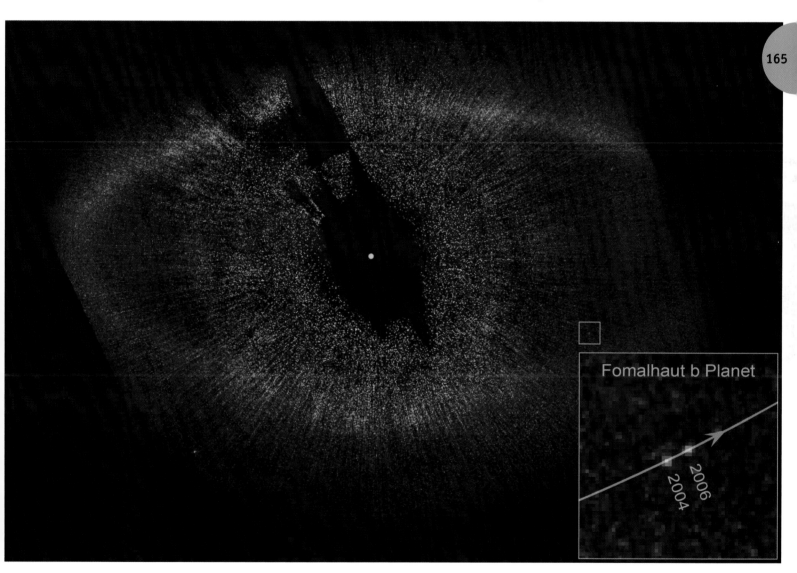

Fomalhaut b Planet

2006
2004

the world and attempt to construct from them the growth of a human being from birth, through adolescence, and into adulthood. From this, we learn that full development of the planets takes place over a period of tens of millions to hundreds of millions of years.

Above: This Hubble Space Telescope image shows the newly discovered planet Fomalhaut b (see the small white box at lower right) orbiting its parent star, Fomalhaut. The inset at bottom right is a composite image showing the planet's position during Hubble observations taken in 2004 and 2006. The white dot in the center of the image marks the star's location; the region here is black because a coronagraph was used to block out the star's bright glare so that the dim planet could be seen. Fomalhaut b is 1 billion times fainter than its star. The red dot at lower left is a background star. This planet is the one predicted by the offset ring seen on the opposite page.
Credit: *NASA, ESA, P. Kalas, J. Graham, E. Chiang, E. Kite (University of California, Berkeley), M. Clampin (NASA Goddard Space Flight Center), M. Fitzgerald (Lawrence Livermore National Laboratory), and K. Stapelfeldt and J. Krist (NASA Jet Propulsion Laboratory).*

the star Upsilon Andromeda (υ And): the first instance of several planets orbiting together around the same star. After the first companion of υ And was found in 1996, another two were discovered. With the case of υ And, a new chapter in planetary science was opened: our planetary system was not alone. There was now at least one other "sun" accompanied by a family of planets, to be sure with somewhat elliptical orbits; this behavior means that the planets would be notably different in temperature during the path covered over an orbit, which is quite unlike the situation of the Sun's planets. Of the three planets of υ And, the intermediate one completes an orbit in 242 days, at a distance from the star comparable to that of Venus from the Sun. The orbit of the outermost planet is at a distance similar to that ranging from the Earth to the region of the asteroids, between Mars and Jupiter.

In 2002, a second extrasolar planetary system was discovered, when the second and third planets were found around the star 55 of the constellation Cancer (55 Cnc). This star is also a G type, with a mass and temperature close to that of the Sun. Up to now this continues to be the system with the largest number of planets — five of them. The fourth was announced in August 2004, with a period of only 2.8 days but, most interestingly, with a minimum mass comparable to that of Neptune, only 14 terrestrial masses.

The periods of these five planets (their "years") range from 2.8 days, to 14, 44, and 260 days, and finally to the maximum of 14.3 years. The planet with the longest period, 55 Cnc d, marked a new stage in exoplanet research because it was the first gas giant planet of the Jupiter type to be found at a distance from its star similar to that of Jupiter to the Sun.

On June 16, 2008, an ESO press release announced the discovering of three "super-Earths" around the star HD 40307; the smallest has a mass of only 4.2 terrestrial masses. Commenting on the news, the fruit of five years of labor, Michel Mayor — the previously mentioned discoverer of 51 Peg b — said, "It is clear that these planets are only the tip of the iceberg. The analysis of all the stars studied with our HARPS spectrograph at the 3.6-m telescope at the La Silla Observatory shows that about a third of the solar-type stars are accompanied by super-Earths or Neptune-type planets. It is very likely, then, that there are planets like the Earth that are as yet beyond the sensitivity of our instruments."

While up to a few years ago we were able to think that ours was the only planetary system, as of the end of 2008 the *Encyclopedia of Extrasolar Planets* listed 35 multiple systems comprising 77 planets (about 25 percent of the known exoplanets). Besides the Cnc 55 system, there is another system of four planets, and four systems with three. And so we have at our disposal a rich laboratory for the study of the formation and evolution of planetary systems. We know now that these systems, with their hot giants very close to the mother star and with planets in

Above: A comparison of two planetary systems: 55 Cancri (top) and our own. Blue lines show the orbits of planets, including the dwarf planet Pluto in our solar system. The 55 Cancri system is currently the closest known analogue to our solar system. Both have similar stars, in mass and age. Both have planetary systems with giant planets in their outer regions. The giant located far away from 55 Cancri is four times the mass of our Jupiter, and completes one orbit every 14 years at a distance of five times that between Earth and the Sun. Both systems also contain inner planets that are less massive than their outer planets. But the planets known so far to orbit 55 Cancri are all larger than Earth, and orbit closer to their star than Earth is to the Sun. Image credit: *NASA/JPL-Caltech.*

orbits that are very elliptical and very inclined, can have a structure completely different from that of our Solar System. Current theories are being put to the test and now need to respond to new questions: How are the masses of the various planets distributed? Can they move from the inner to the outer parts of their system, or vice-versa? And are the planes of their orbits generally coplanar, as is the case in our Solar System? How did the Hot Jupiters get so close to their mother stars? Did they form at a greater distance away and then migrate there? The newly observed data give us new directions for elaborating new theories. A system like ours, with more or less circular orbits, may be the exception rather than the rule.

It is reasonable to think that any number of the gas giant planets would have their "moons," just as Jupiter and Saturn have theirs. And we emphasize that, just like the satellites around Jupiter and Saturn (for example, Titan, that object of the extraordinary Cassini-Huygens mission to Saturn, and Europa, the satellite of Jupiter) are the focus of current research looking to discover environments in our planetary system that could be compatible with the presence of life-forms (see p. 147).

How Do We Define a Planet Now?

This question may appear naïve, and its answer obvious. A few years ago you could simply say, "A planet is one of the nine large bodies that do not shine by their own light and which orbit around the Sun." Today the situation has changed, and the question has sparked a considerable controversy. The introduction onto the scene of extrasolar planets has reopened the discussion as to what qualities should be the characteristics that allow a

celestial body to be given the title of "planet."

In the General Assembly of the IAU at Prague in August 2006, the majority of the assembled astronomers decided that Pluto, with a mass much smaller and an orbit far more elliptical and inclined than that of the other solar planets, should be demoted from the classical list of planets in our Solar System and inserted into the new category of "dwarf planet": a

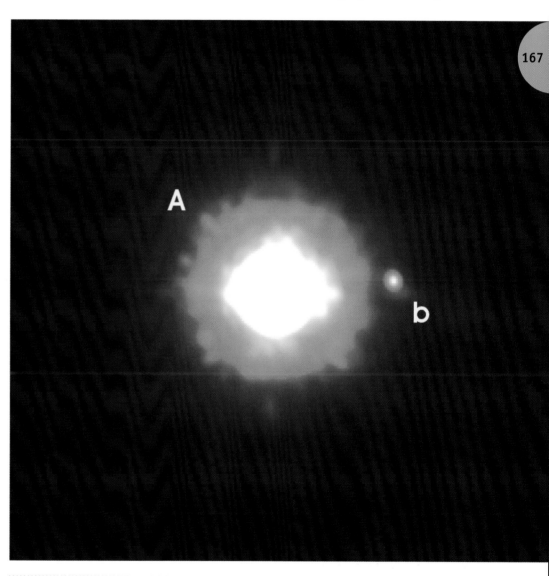

Above: This infrared image from the Very Large Telescope (VLT) at the European Southern Observatory (ESO) in Chile shows a feeble point of light to the right of the star GQ Lupi; this is its cold planet. It is 250 times fainter than the star itself and orbits at a distance of roughly 100 astronomical units from its star.

celestial object in orbit around the Sun with a mass that, while sufficient to pull it into a roughly spherical shape, is still not massive enough to clear out other bodies from the vicinity of its orbit and gravitationally dominate its region of space. This last characteristic is considered necessary for any body

communication of their discoveries. (In June 2008, the executive committee of the IAU determined that any dwarf planet beyond Neptune, like Pluto, will be called a "plutoid" in its honor.)

For an exoplanet, the definition given by the appropriate IAU Working Group requires that it be a body that has a mass less than that which would allow the thermonuclear fusion of deuterium (such a mass is normally calcu-

lated as 13 Jupiter masses) in orbit about a star or a stellar remnant.

This definition of an extrasolar planet puts into play four particular extrasolar bodies that constitute a separate category with respect to the others. They are objects with a few terrestrial masses, three of them found to orbit around the pulsar PSR 1257+12. They were actually the first of the extrasolar planets to be discovered; the Polish astronomer Alexander Wolszczan

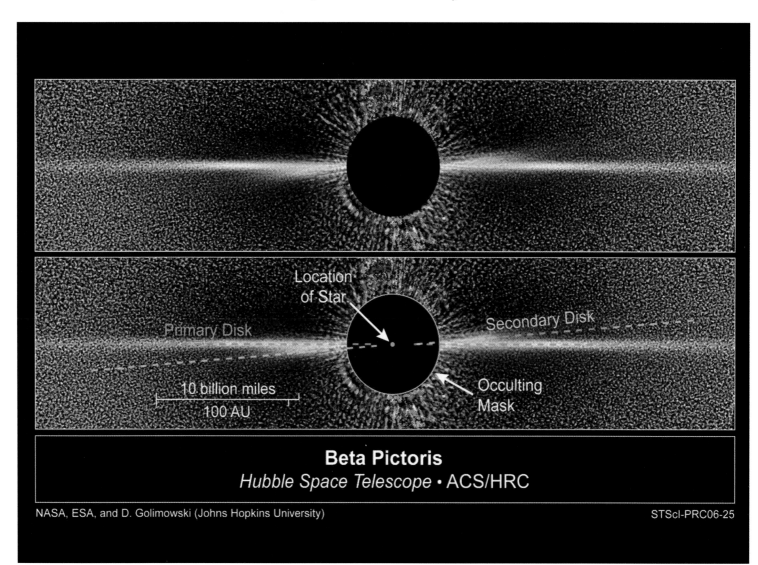

Beta Pictoris
Hubble Space Telescope • ACS/HRC

NASA, ESA, and D. Golimowski (Johns Hopkins University)   STScI-PRC06-25

to retain membership in the classical category of solar planets. It is understood that this definition, though somewhat arbitrary, is still necessary because researchers need to have at their disposal a common terminology that permits them to categorize these objects and avoid confusion in the

Above: This Hubble Space Telescope view of Beta Pictoris shows a primary dust disk and a much fainter secondary dust disk, circumstantial evidence for the existence of a planet in a similarly inclined orbit. Astronomers used the Advanced Camera's coronagraph to block out the light from the bright star (the black circle in the center of the image).
Credit: *NASA, ESA, D. Golimowski (Johns Hopkins University), D. Ardila (IPAC), J. Krist (JPL), M. Clampin (GSFC), H. Ford (JHU), and G. Illingworth (UCO/Lick) and the ACS Science Team.*

detected their presence with the Arecibo radio telescope in 1991. Pulsars are rapidly rotating, super-dense neutron stars, the product of supernova explosions and emitters of regular pulses of intense radiation. (The presence of a planet modifies the frequency of their signal in a periodic way.) It is still not clear how such formidable explosions, which destroy the mother star, could still leave behind nearby planets.

Just beyond the 13 Jupiter mass limit is the category of "brown dwarfs." These are the smallest stars, intermediate between solar-type stars and giant planets. They are intermediate objects, because the temperature in their center is not sufficient to ignite the thermonuclear reaction that burns hydrogen into helium, but only a brief reaction requiring deuterium. Thus, they are lower-mass stars, which, in their evolution, remain too cold to be seats of stable and prolonged nuclear reaction. The greater part of the new exoplanets have masses near this deuterium-burning limit, and it is sometimes rather difficult to distinguish them from brown dwarfs because the Doppler method of radial velocities (which has found more than 90 percent of the exoplanets up to now) cannot determine, by itself, the true mass of the planet, but only a lower limit for that mass. This method, from the displacement in the parent star's spectral lines due to the Doppler effect, deduces the presence of a planet from the variations of the speed of the star in its movement towards or away from Earth.

Other methods for discovering planets are coming to our assistance, however, enriching the catalog of extrasolar planets and resolving the problem of minimum masses. Among these techniques are observing the slight dimming of a star's light when a planet crosses in front of the planet (*transits*) in our line of sight (some 50 planets have been discovered with this transit method); very precise astrometry using space telescopes that can measure the wobble of a star; and *gravitational microlensing*, which occurs when the gravitational field of a star acts like a lens, magnifying the light of a distant background star and its planets.

All of these methods have detected the presence of candidate extrasolar planets only in an indirect way. But in September 2004, a team from ESA presented the first direct image of a planet, 2M1207b, which was subsequently confirmed by the Hubble Space Telescope. In November 2008, that same instrument obtained the first extraordinary image in visible light of the planet Fomalhaut b, orbiting its mother star every 872 years. Its sun is a young (200 million years old) star, a mere 25 light-years from us, the brightest star in the constellation Piscis Austrinus (The Southern Fish). As of this writing, the catalog of extrasolar planets lists 11 that have been imaged directly.

How Do We Make Sense of Our "Collection" of Planets?

Because of the enormous distances, we can't talk of actually exploring these places; that's for the far distant future. But we can begin to draw some interesting inferences from the data at hand already. It's the normal way that science proceeds: after you collect your experimental data, you classify it into various groups and come up with various hypotheses or theories that are consistent with the data. We are beginning to pass into that phase in the study of extrasolar planets, making the first tentative passes at an analysis of their characteristics.

Our collection of 333 extrasolar planets is hardly complete; in the next 10 years we're likely to find hundreds or thousands more examples. Those are the kinds of numbers needed to make sense out of all the different varieties of extrasolar planets. And as more lower-mass planets appear, we should be ready for more surprises.

Astronomers are not interested just in finding new planets; they want to bring to light and find statistical relationships among their various properties, to indicate the next promising direction for future research. They are looking for correlations between the masses of the planets, their sizes, the periods of their revolutions, and other characteristics of their orbits. For planets observed to transit their mother stars, so that an estimate of the planet's dimensions can be made, you can add to the mix the density, which is an indication of its internal composition. By observing planet transits, you can also analyze their atmospheres spectroscopically. For all the planets, you can measure the spectra of the mother stars and look for a correlation between the "metal" content of the stars and the probability that a star has planets. All of these ingredients are needed before one can come up with a hypothesis to make sense of this collection of data.

As we increase the number of known planetary systems, we have completely revised the study of the evolution of protoplanetary disks and the dynamics of the probable "migration" of planets during the formation of such systems. The study of extrasolar systems has significantly deepened our understanding of our solar system. It has given rise to the debate among scientists as to which characteristics of our Solar System are commonplace enough, and which are quite rare.

Future Developments

Many programs involving the largest Earth-based telescopes are now aiming to obtain direct images of giant extrasolar planets. Meanwhile, projects are underway to monitor hundreds of thousands of stars for possible transits. The COROT (France/ESA) satellite, in orbit since December 2006, plans on monitoring the light curves of 12,000 stars in each series of observations, and it has already discovered some exoplanets. Starting in April 2009, the Kepler space mission will be dedicated to a four-year mission to search for Earth-like "twins" by looking for transiting planets, similar to our own, orbiting about 100,000 stars in our Galaxy in the constellations Lyra and Cygnus. If medium- and small-sized planets are common, Kepler could find hundreds of them, of which a few dozen may be "twins" of Earth in habitable zones around their stars.

In 2013, if all goes according to plan, the next generation James Webb Space Telescope will be launched, fruit of a collaboration between ESA and NASA, with a 6.5-meter mirror. JWST will orbit

at 1.5 million kilometers from the Earth (much more than the 500-kilometer altitude of Hubble). While looking at other galaxies in the primitive (distant) universe, JWST will actually study planetary systems in formation. It should be able to directly image planets and give us information about the chemical composition of their atmospheres.

Proceeding from that will come the search for *biological markers*, such as oxygen, ozone, water vapor, and methane, and even indications of photosynthesis in their spectra, which could signal the possible presence, now or in the past, of biological agents that on Earth would be indications of life on their surfaces. The resulting information will contribute to the planning of missions planned for the coming years such as TPF (Terrestrial Planet Finder) and Darwin from NASA and ESA. The preliminary studies for these missions propose a battery of five or six telescopes, each a meter and a half in diameter. They will "fly in formation" in an orbit a million and a half kilometers from the Earth, four times more distant than the Moon, to look for terrestrial-type planets using *interferometry*, where the light from each telescope is compared against the next in order to reduce the effect of the brightness of the mother star by a factor of 10 million and thus make any planets easier to see.

The Significance of the Discovery of Extrasolar Planets

The advent of extrasolar planets has opened a new page in astronomy, and it has already given a formidable impulse into both theoretical and experimental research. Beyond the scientific results, it is clear that these discoveries have also relaunched the question of the presence of other habitable worlds. The search for Earth's twins and the

possibility of life on them remains a priority with both a scientific and an emotional resonance.

Fundamentally, the study of astronomy is not merely using a telescope or a computer. While one needs exceptional patience to record the observational data — ever more refined, thanks to the ever-advancing technology — and to come up with ever more comprehensive and complete hypotheses, astronomy goes deeper; it nurtures our desire to comprehend how our brief and fragile existence can be connected with that of the planets, stars, and galaxies. Like a six-year-old child unable to tell a star from a planet, we can let our imaginations go and ask ourselves the question if there is, in some other world, someone we can play with.

For millennia, humanity has posed this question. Epicurus, in his letter to Herodotus, wrote: "There are an infinite number of worlds, some like ours, others diverse ... nothing hinders such an infinity ... and we shouldn't think that such worlds necessarily all have the same form. No one can prove that ... in such a kind of world it could not be possible that there are present the seeds from which can arise animals and plants and all the rest of the things that we see."

But now, after centuries of anticipation and often fantastic speculation, we have turned the page; we have sure, reproducible evidence that many planets actually exist in the known universe. Still, we need to distinguish the possibility of the existence of an embryonic form of life on the planets from the positive affirmation that certain exoplanets have, and continue to have over time, the necessary conditions for primitive forms to evolve to the point of consciousness and intelligent life.

At this point, it is worth looking at the theological implications connected with the eventual discovery of intelligent life in a plurality of worlds in our universe. In 1277, the bishop of Paris, Stephan Tempier, condemned the proposition of traditional Aristoteleanism according to which the Prime Cause could not have made other worlds (arguing that to hold such a belief diminished God's omnipotence). But when Thomas Aquinas supported the existence of only one world, he deemed that such uniqueness was necessary to affirm the uniqueness of the Creator (S.Th I, q 47, ad 3). Still, the medieval concept of "many worlds" does not really correspond to our present discussion of many extrasolar planets; rather it is somewhat more related to recent cosmological speculations about "multi-universes."

In more recent times, we can recall the openness of Fr. Angelo Secchi, S.J., (a pioneer of astrophysics) concerning the possibility of extraterrestrial life; likewise that of the Barnabite priest, Fr. Francesco Denza, first director of the reformed Vatican Observatory. And we note the manifest confidence of Pierre Teilhard de Chardin, S.J., who wrote in 1953, *"Une suite au problème des origines humaines: La multiplicité des mondes habités"* ("An aspect of the problem of the origin of humanity: the multiplicity of habitable worlds").

Among the astronomers at the Vatican Observatory, there have been diverse opinions on the possibility of eventually encountering ET. But I have the impression that these are determined not so much by the evidence of the research data itself, as by the

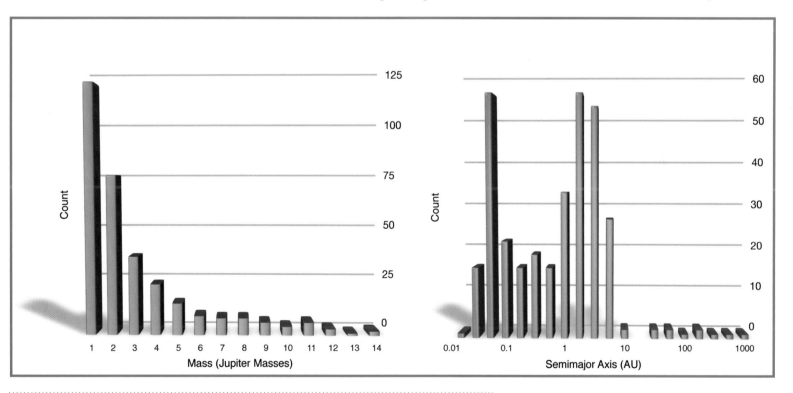

Above: The range of masses (left) and average orbital distances (right) of the 333 exoplanets discovered as of the end of 2008. (One AU is the average distance from the Earth to the Sun.) Most discovery techniques have an easier time finding planets with larger masses and planets closer to their stars, but more recent discoveries and newer methods have begun to reveal planets in a wide range of sizes and positions. (Data from the online Extrasolar Planets Encyclopedia.*)*

general sense of how the universe and life works, with the basic "Copernican" sense that there are no special privileges for planet Earth.

I believe that certainly patient and methodical observations are to be preferred to the speculations of science fiction writers. While it seems reasonable to suppose that there is a strong possibility the universe will come up with new forms of life every time conditions allow for it, one's imagination may run more freely than one's science. One must remember that the possibility of entering into contact with an evolved alien form of life is terribly limited. If, 20 years from now, thanks to future space telescopes, we are able to observe planets with the appropriate characteristics and if, for example, in their atmospheres there is a strong concentration of oxygen or some other indication of biological activity that would support the probable presence of life, we would still remain quite frustrated. From the point of view of philosophy, such a discovery would be extremely important, but we would still not in fact be able to tell what sort of life we were dealing with. Still, it is to be expected that a scientist will focus on what can be obtained with the available means today. While the goal is admittedly distant, the field of investigation is fascinating and rich with promise.

We can say that the discovery and the initial characterization of extrasolar planets would let us broaden our understanding of the cosmos, moving from the realm of stars — which we have long studied — to the more intimate scale of "worlds," which are instinctively more attractive to us ... the kinds of places where human beings can have adventures. Certainly, the planets discovered up to now are not likely to be hospitable for life-forms, but we are gradually approaching the discovery of bodies with ever smaller masses; there's no reason to think that we won't eventually come across one similar to our Earth.

The great undertakings of science — and such is the study of extrasolar planets, which has involved such a large number of international research groups, both theoretical and experimental — have more than just practical benefits, such as improving our technology. It engages the human mind and heart with its untiring search for truth and beauty. It is encouraging to see how many people of science have dedicated their energies to a project whose scale is such that only their successors will be able to see its fruit. In June 2007, the Vatican Observatory dedicated its 11th summer school to introducing more than two dozen young students to this fascinating interdisciplinary topic.

The founder of the Society of Jesus, Ignatius of Loyola, in his brief auto-biographical note written (in the third person) at the request of his first companions, left us this description of his early and decisive time of conversion: "His greatest consolation came from the contemplation of the heavens and the stars, which he would gaze at long and often, because from them there was born in him the strongest impulse to serve Our Savior." What he experienced was not just a delicate sensibility and an aesthetic sense, but an interior movement that allowed him to discern the loving action of God unfolding in creation and gave him the desire to participate in it. Such contemplation nourishes the vitality of faith.

This is well expressed in Sacred Scripture: "[He] sends forth the light, and it goes; He called it, and it obeyed Him, trembling; the stars shone in their watches, and were glad; He called them, and they said, 'Here we are!' They shone with gladness for Him who made them" (Baruch 3:33-34).

And again: "When I look at Your heavens, the work of Your fingers, the moon and the stars that You have established; what are human beings that You are mindful of them, mortals that You care for them?" (Psalm 8:3-4). ●

FR. GIUSEPPE KOCH, S.J., a physicist, is the superior of the Vatican Observatory Jesuit Community in Castel Gandolfo and vice director for administration. This chapter was translated by Guy Consolmagno.

PONDERING STARS

Above: *The Hubble Ultra Deep Field, a composite based on images taken from late 2003 through January 2004, contains an estimated 10,000 galaxies. Aiming at a relatively empty part of the sky in the southern constellation Fornax, the goal was to look past nearby brighter objects to image galaxies at a great distance, and hence representative of material in the earliest epochs of the universe, to within the first billion years after the Big Bang.* **Credit:** *ESA and NASA, Hubble Space Telescope.*

IS BIG BANG COSMOLOGY IN CONFLICT WITH DIVINE CREATION?

• Fr. WILLIAM R. STOEGER, S.J. •

What Is the Big Bang?

There is compelling evidence that our universe emerged from an extremely hot, dense primordial state about 14 billion years ago — the Planck era, which is often considered the direct result of the Big Bang. From that fiery epoch it has gradually expanded and cooled. And as it cooled, it has become more and more lumpy, and more and more complex. As ever-lower temperatures were reached, simpler, more basic entities and systems combined and formed an ever more complex and diverse array of evolving systems — particularly in cooler, more protected, more chemically rich environments.

But what is the Big Bang? Strictly speaking, it is the past limit of the hotter, denser phases we encounter as we go back further into the history of the universe. Not only is it observationally inaccessible, but it also lies outside the reliability of the classical (non-quantum) cosmological models upon which we depend.

This does not mean that there

Right: Fr. Georges Lemaître with his mentor Sir Arthur Eddington in 1938.
Credit: "Archives Lemaître"
Université catholique de Louvain.
Institut d'Astronomie et de Géophysique
G. Lemaître. Louvain-la-Neuve. Belgique.

is no evidence supporting Big Bang models. There is pervasive compelling evidence from a number of independent quarters — most notably the cosmic microwave background radiation, the universality of large systematic redshifts of distant galaxies, and the abundances of deuterium, helium, and lithium. There is no doubt that, as we go back further and further into the past, the universe was hotter and hotter, and denser and denser. The observational inaccessibility is of the earliest hotter denser phases, and of whatever event or state triggered the universe's expansion and cooling.

What will quantum cosmology be able to tell us about it? By considering the recent educated scientific speculation on what may have led to the Big Bang and the Planck era, we shall find that quantum cosmology — and the physics upon which it relies — promises to reveal a great deal, but cannot provide an alternative to the traditional philosophical notion of divine creation, creation from nothing, in accounting for the universe's ultimate origin. Any understanding it might provide, no matter how physically fundamental, will require a deeper explanation or basis for its existence, order, and properties. In other words, it will not be self-subsistent or self-explanatory. But at the same time, quantum cosmology indirectly poses these ultimate questions, which it cannot answer, and in so doing, points towards — and is consonant with — divine creation.

The Planck Era and "the Beginning" of the Universe

But what about the Planck era and the Big Bang? What generated this extreme primordial state? Was this the very beginning of the universe?

If we go all the way back in time

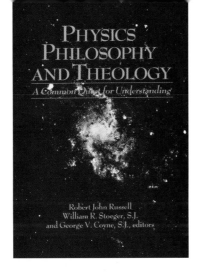

as far as we can go, we find that at a certain point our standard model of the universe describes it as having infinite temperature and infinite density. In the version that best fits what our universe is like, this point, which is often referred to as "the initial singularity" or "the Big Bang," would have occurred about 13.7 billion years ago — but only the tiniest fraction of a second before the universe was at the temperature of the Planck era, 10^{32} K.

There is a problem, however, in taking this "initial singularity" or "Big Bang" point seriously. The fact that it involves infinite temperature and infinite density serves as a warning that this did not actually happen. It is simply a "prediction" of the model which does not represent what really occurred. In fact, there are very strong indications that

Quantum cosmology indirectly poses ultimate questions which it cannot answer....

the key assumptions upon which this very reliable model is based — those of Einstein's theory of gravitation, general relativity — break down when the universe is at or above the Planck-era temperature. The model is very reliable below that temperature, but severely fails in describing the physics and behavior during the Planck era itself, or during any era preceding it.

From this discussion we can clearly see, then, that the Big Bang, or even the Planck era, is not "the very beginning" of the universe. It certainly is "the beginning" according our provisional models of the universe. But those models are completely inadequate precisely in the region of the Big Bang! Thus, on the basis of what we know so far, we can say very little about the Big Bang and Planck era, or about what generated them. As we shall see, however, research in quantum cosmology — though not yet yielding complete and reliable results — has begun to shed some significant light upon some possibilities and some of the characteristics of that primordial cosmic state.

Eventually, from a philosophical point of view, we shall want to determine whether or not an adequate theory of the conditions in the Planck era, and a reliable account of what led up to it, will ever be able to model an "absolute beginning" of the universe. Or more fundamentally, are physics and cosmology capable of providing an ultimate explanation for the universe and its principal features? If so, then they would be viable alternatives to the philosophical *creatio ex nihilo* idea of creation, which constitutes the basis for the theological doctrine of creation in Judaism, Christianity, and Islam. If not, then they would be complementary, and not on a level equivalent to that of philosophical or theological explanation. But before we delve more deeply in that issue, we look briefly at what quantum cosmology is suggesting.

Insights from Quantum Cosmology

Why does the basic physics underlying the standard cosmological models, and their modifications (e.g., the addition of an inflationary epoch), break down at the very high temperatures, or energies, which characterize the Planck era? One simple answer is to say that the universe is too hot for space and time (or more correctly space-time) to exist as smooth continua. The fluctuations in geometry are so large that the concept of space-time as we usually model it — as a smooth, connected manifold — is no longer valid. Instead we have to find an adequate way of representing this highly energetic state with a discrete, broken-up, foamlike structure, which becomes space-time when the temperature falls below 10^{32} K, and the universe emerges from the Planck epoch. In other words, we need a quantum description of space-time and therefore of gravity. This is because the basic physics of space and time is intimately linked with the gravitational field, which in turn is determined by the mass-energy distribution throughout the universe. (Remember mass and energy are equivalent: $E = mc^2$!) Mass-energy generates gravity and therefore space-time, but space-time and gravity in turn tell mass-energy how to move. We thus need to somehow marry Einstein's gravitational theory, which wonderfully and accurately describes this fundamental link, with quantum theory, which deals with the particle-and-wave-like character of reality at submicroscopic levels. So far this challenge has proved extremely difficult, and has not been met.

Over the past 40 years or so, there have been a number of different approaches that have been taken towards developing a fully reliable theory of quantum gravity which could be used by quantum cosmology to describe the very early universe, and therefore the Planck era and what may have possibly preceded it or led to it. At present, the most highly developed and promising of these are superstrings, loop quantum gravity, and noncommutative geometry. I shall not spend time here describing these fascinating and tantalizing ideas in any detail. There is just not space here for that. (For a readable introduction to these ideas in this context, see Lee Smolin's book *Three Roads to*

... the Hartle-Hawking proposal has been touted by some as indicating that physics and cosmology now can provide a universe emerging from nothing.... However, this is an illusion, at least from a philosophical point of view.

Quantum Gravity, published by Basic Books in 2001.) Here I shall just mention some of the other consequences, or preliminary conclusions, such theories suggest for quantum cosmology.

But first we must realize there have been a number of significant contributions to quantum cosmology apart from those directly connected with developing a full and adequate theory of quantum gravity, or a fully unified theory of all four fundamental interactions. These have been what might be called semi-classical quantum cosmology treatments. These incorporate some of the basic insights from standard quantum theory and from Einstein's general relativity and may indicate some of the key features that characterize the Planck era and what led to it. (As Chris Isham has emphasized, "Certain general properties are expected to hold in any quantum gravity theory....") These approaches are enabled by the requirement that any quantum cosmology or quantum theory of gravity must, as the temperature or energy decreases, yield the reliable classical, or non-quantum, models we already have — like the standard cosmological model. Working the other way, we can construct the semiclassical quantum versions of these classical models and see what sort of quantum corrections are expected to occur as we go to slightly higher temperatures which trigger the transitions to the quantum-cosmological regime.

Among some of the key people who contributed to such approaches are John Archibald Wheeler, Bryce DeWitt, Stephen Hawking and James Hartle, and Alex Vilenkin. Wheeler and DeWitt formulated the very elegant and suggestive Wheeler-DeWitt equation, which describes in simple terms the behavior of the quantum-mechanical "wave-function of the universe." This cosmic wave function would under certain conditions have a definite probability of issuing in our classical universe, which would then expand and cool, as general relativity and the standard universe models prescribe. It is important to realize that the Wheeler-DeWitt equation does not contain time as such. In the purely quantum regime, the wave function of the universe in some sense "just is." However, there is a sense in which time can emerge from the Wheeler-DeWitt equation as the transition between the wave-function of the universe and the classical universe itself occurs. Hartle and Hawking later extended this work, and showed, by using the concept of imaginary time — by which one treats time exactly like a spatial dimension — and by conceiving that there is no initial three-dimensional spatial boundary to the universe, that we can in a consistent way obtain from

Physics can never tell us how we get from absolutely nothing to something.

the cosmic wave-function a universe like the one we inhabit. They also obtain a very early inflationary phase for this universe, which seems to be required. This is called the "Hartle-Hawking no-boundary proposal" for the origin of the universe, an amazing idea and result, but one which depends on a number of assumptions that are not easy to justify.

(For a nontechnical description, see C. J. Isham's article "Creation of the Universe as a Quantum Process," in *Physics, Philosophy and Theology: A Common Quest for Understanding*, edited by Robert John Russell, William R. Stoeger, S. J., and George V. Coyne, S. J., and published by the Vatican Observatory in 1988.)

It is notable that the Hartle-Hawking proposal has been touted by some as indicating that physics and cosmology now can provide a universe emerging from "nothing." Because there is no boundary, nor any classical time, that can be defined, there

is a sense in which the physics seems to indicate that it "just appears from nothing." However, this is an illusion, at least from a philosophical point of view. At the very least, one needs the existence of the wave-function of the universe and the ordered behavior described by the Wheeler-DeWitt equation itself. Where did these come from, or why are they as they are, rather than as something else?

Furthermore, as M. Bojowald and H. A. Morales-Técotl have pointed out, this proposal really does not eliminate the singularity, which they accept as "a point of creation." This is clear because the wave-function does not vanish at the singularity. The Wheeler-DeWitt equation, and proposals for solving it, like that of Hartle and Hawking, and Vilenkin, present simplified models, or descriptions, of some features we might expect from the quantization of the gravitational field, and of the early universe. But they are by no means adequate. They should approximate what happens far from the singularity, but they certainly are incapable of describing what happens near it. Much less do they describe "the process" by which the creation of the universe took place, understood in the radical philosophical sense.

String theory has recently generated two other popular but still inadequate scenarios for triggering the Big Bang and providing a possible way of understanding the emergence of the universe from the Planck era. One is "the pre-big-bang scenario," and the other is "the ekpyrotic scenario." Because of symmetries in string theory, including time-reversal and what is called "T-duality," two completely different phases of the universe are allowed — a pre-big-bang phase, in which the universe collapses from an almost empty state an infinite time ago, to become very dense and very hot leading to the Planck era. However, like many other quantum gravity theories, string theory does not allow a singularity — the volume has a minimum, and the density, temperature, and curvature have maxima. When these are reached, the universe bounces and enters the post-big-bang phase. Thus, according to this suggestion, it is very clear that the Planck era and the Big Bang are not the beginning of the universe, nor even of time. One of the principal difficulties with this suggestion is that there is not yet a satisfactory account of how the transition (the bounce!) from one phase to the other may have occurred. In the ekpyrotic scenario, our universe is simply one of many large membranes (D-branes) floating in a higher dimension space. These "branes" are a natural consequence of string theory. Periodically, because of the gravitational attraction between them, these branes collide with one another, triggering a big-bang-like event. However, not any pair of colliding branes will yield the Big Bang and the universe we have. For two branes to do that would require that the branes themselves and the collision between them be "finely tuned" or carefully orchestrated. For instance, the branes involved should be almost exactly parallel. (I am indebted to George Ellis, the well-known cosmologist from the University of Cape Town, for pointing out this qualification to me.)

A general and fundamental conclusion we can draw from our brief discussion of quantum cosmology is: Any more reliable scenario for the origin of the Planck era, and the triggering of the expansion and cooling of the universe from that state, requires other detailed physics describing some physical structure or states that in some sense underlie or explain the Planck era itself. Any such account will always demand some further explanation or physical foundation — and ultimately an adequate metaphysical foundation or ground. As George Ellis has remarked, quantum cosmology assumes that all of the structure of quantum field theory, superstrings, or other organizing structures, pre-exists the universe itself, since they determine its emergence. So, where does all of this structure reside? And how does it trigger the coming into being of the whole physical universe? We might also ask with Ellis, where do these structures reside after the universe has emerged? These questions push us beyond where the natural sciences, or perhaps any human inquiry, are able to take us.

Physics as such can specify in great qualitative and quantitative detail how we get from one physical state to another, or what the underlying constituents or factors of a given state are. It can do this if it has adequately modeled the regularities and relationships involved. However, it cannot in principle account ultimately for their existence or for the particular form those structures, regularities, and relationships take. To put this in temporal terms, which are not essential to the issue, physics can never tell us how we get from absolutely nothing — no space or time, no matter or energy, no wavefunction or field, nothing physical at all — to something that has a particular order. There is no physics of "absolutely nothing." Thus, though physics can shed a great deal of light on many other questions having to do with the universe, it evidently cannot help us in illuminating the ultimate ground of order or of being. This is precisely why physics in general and quantum cosmology in particular do not provide an alternative account of the creation of the universe, philosophically or theologically speaking.

And so now we turn to discuss the philosophical concept of "creation from nothing" — *creatio ex nihilo* — as a complementary, not an alternative, understanding of the origin of the universe, and of reality in general.

> *The Creator empowers the physical processes to be what they are. The Creator does not replace them.*

The Basic Insight
of *Creatio ex Nihilo*

The basic reason why *creatio ex nihilo* is complementary to any scientific explanation, including whatever quantum cosmology theoretically and observationally reveals about the "earliest" stages of our universe — or multiverse — and not an alternative, is that it does not and cannot substitute for whatever the sciences discover about origins. It simply provides an explanation or ground for the existence and basic order of whatever the sciences reveal. The Creator empowers or enables the physical processes — including whatever primordial originating processes and entities, whatever they are — to be what they are. The Creator does not replace them. Nor, as we have just seen above, can what quantum cosmology discovers and models substitute for what *creatio ex nihilo* accomplishes — that is, providing an ultimate ground of existence and order. In our discussion at the end of the previous section, we found rather strong indications that any physical process or dynamical structure that would account for, or generate, the extreme conditions marking the Planck era, or triggering "the Big Bang," requires a more fundamental physical explanation or grounding. Nothing we are familiar with in the physical or biological worlds — or in reality generally — stands on its own without requiring some cause and context. Nothing we can investigate scientifically completely explains its own existence and characteristics. Thus, whatever we find in

Creatio ex nihilo is not an answer to the question of temporal origin.... Creatio ex nihilo is, instead, about the ultimate ontological origin of reality....

quantum cosmology will always raise further questions for understanding. Thus, an infinite regress of questions of physical origin is inevitable. And no member of this chain of origins, nor the entire chain itself — even if its infinity were realized — would provide an ultimate grounding for existence and order.

What *creatio ex nihilo* provides, then, is an ultimate ground of existence and order for the universe — and for reality as a whole. It does this by proposing a self-subsisting, self-explanatory "cause" — the Creator — which is the fundamental source of being and order, and in which all existing things participate. As such, this ultimate ground of being and order is not another entity or process in the universe, which can be discerned or isolated from other physical causal factors and entities. It is not scientifically accessible! And yet, it is causally distinct from them, because, without it, nothing would exist. As such, it does not substitute for created causes — it endows them with existence and efficacy. One way of putting it is that this Creator, however we attempt to describe it, is the necessary condition for everything, and

the sufficient condition for nothing. Events and changes occur, and entities and systems emerge and subside into their components, only through the created, or "secondary causes" which the Creating Primary Cause sustains. In fact, the rich philosophical tradition shared by Judaism, Christianity, and Islam uses the complementary categories primary cause and secondary cause in just this way.

(Many people have written about *creatio ex nihilo*. For brief summary treatments, you might want to look into Catherine Mowry LaCugna's *God for Us: The Trinity and Christian Life* (published by Harper-San Francisco in 1993); Langdon Gilkey, "Creation, Being, and Nonbeing," in *God and Creation: An Ecumenical Symposium*, edited by David B. Burrell and Bernard McGinn, and published by the University of Notre Dame Press in 1990; or my own articles, "The Origin of the Universe in Science and Religion," in the book *Cosmos, Bios, Theos: Scientists Reflect on Science, God, and the Origins of the Universe, Life and Homo Sapiens*, edited by Henry Margenau and Roy A. Varghese, and published by Open Court in 1992; and "Conceiving Divine Action in a Dynamic Universe," in *Scientific Perspectives on Divine Action: Twenty Years of Challenge and Progress*, edited by Robert John Russell, Nancey Murphy, and myself, and published just this past year by Vatican Observatory Publications and the Center for Theology and the Natural Sciences.)

Presuming for the moment that there are no serious reasons for dismissing the basic concept of *creatio ex nihilo*, how can we understand it better?

First, it is crucial to realize that when we talk about God, or "the Creator," we will never be able to have an adequate concept of that. It will always be beyond us — radically transcendent. But at the same time, we can point to the mystery of existence and order at the depths of reality and of our experience, and say something very tentative about creation and what it requires. There will be some ways of speaking about God and God's creative action that are less inadequate than others! In the same vein, we have to acknowledge that, when we talk about God "causing" or "acting" when God creates, we are speaking metaphorically or analogically. God acts or causes in a very different way than anything in our experience acts or causes. And yet there is some legitimate content to those assertions, in the sense that God somehow endows things with existence and with their specific being in the ultimate sense, but also through the action of other created causes, which God also holds in existence. Without God, they would not exist! Thus, God as Primary Cause is a cause unlike any other cause — unlike the created, or secondary causes, which God sustains and enables. God is their necessary condition.

Secondly, *creatio ex nihilo* is not primarily an answer to the question of temporal origin. It's an open philosophical question whether or not there was something like a temporal beginning to creation — a first moment, as it were. Certainly, as we have already seen, quantum cosmology points to the separation of the first moment of time as we know it from the origin of the universe itself, if there was one. *Creatio ex nihilo* is, instead, about the ultimate ontological origin of reality — most fundamentally it describes in a very bald and unadorned way the ultimate dependence of everything on the Creator. It is not about a creation event, but about a relationship which everything that exists has with the Creator (as noted by LaCugna). So *creatio ex nihilo* is also *creatio continua*, continuing creation. The relationship between the Creator and the created continues as long as something exists. The Creator sustains or conserves reality — and the universe — in existence. Without God, it would not exist. It has been helpful to conceive the relationship of creation as a participation in the being of the Creator. In this regard, it also seems clear that it is better to conceive the Creator more like a verb, than like an entity. In some ways, the Creator is pure, self-subsisting being, activity, or creativity, in which all things participate. Some philosophers and theologians have traditionally referred to God as "Pure Act."

Thirdly, it is also critical, as we have already implied, to avoid conceiving the Creator as controlling creation, or as intervening in its dynamisms. God, instead, enables and empowers creation to be what it is — and both ultimately endows and supports all the processes, regularities, and processes of nature with their autonomous properties and capacities for activity. Thus, God as Creator does not substitute for, interfere with, countermand, or micromanage the laws of nature. They possess their own integrity and adequacy, which God establishes and respects.

Fourthly, it is often claimed that God as Creator, though transcendent, is immanent in creation and in its activity. Though God as Creator does not function within nature or history as another created (secondary) cause, God is present and active in and through the whole network of processes and relationships, precisely because God is sustaining them and enabling them. We can better understand this by pointing out that transcendence is not about being above and beyond creation as detached from it, but rather being free from any barriers, limitations, or obstructions. Thus, there is no barrier to the ground of being and order being immanent — deeply present and active, but present and active as Creator, not as another created cause — within all aspects of creation. Transcendence does not impede or contradict immanence — it enables it!

Fifthly, the relationship of ultimate dependence and creative immanence is

Quantum cosmology and creatio ex nihilo *contribute deeply complementary and consonant levels of understanding of the reality in which we are immersed....*

not uniform, but instead is highly differentiated — that is, it is different with respect to each entity, organism, system, person, and process. God sustains them all in being, but God is sustaining different things in being, with different properties, capacities, and individualities — and through different constitutive relationships with the world around them. And each responds to its environment and to the situation within which it finds itself — and therefore to God — in different ways.

There is much more that could be discussed about *creatio ex nihilo* and how it is to be coherently understood. But what I have presented here captures the essence of the approach in a way that helps us appreciate the fundamental question it attempts to answer, and why, if properly understood, it cannot be in competition with cosmology or the other natural sciences in explaining the origins of the universe, or of anything emerging within it.

... if properly understood, it cannot be in competition with cosmology or the other natural sciences in explaining the origins of the universe.

Conclusion

Now that we have looked carefully at the way physics, cosmology, and quantum cosmology probe the origins of the universe and the objects and systems that emerge within it, and explored the essential contents and limitations of *creatio ex nihilo*, we can see more clearly how different they are from one another. In particular, we begin to appreciate the detailed scenarios that quantum cosmology constructs and tests, as well as the need to find a physical explanation for any stage of cosmic development, no matter how primordial.

By their very nature, physics and cosmology, as do the other sciences, will always focus on how we get to a particular outcome from another physical configuration by some transition, process, or change. Thus, they attempt to describe in qualitative and quantitative detail "the first" configuration and the physics that enables the transition to the outcome in question. This has proved extremely powerful. However, it has the limitation that it can never deal with the essential ground of being and order, upon which all else rests. *Creatio ex nihilo* as a philosophical — not a scientific — approach attempts to do that. Properly applied, it is not about changes, processes, or transitions — it does not, and cannot, substitute for anything that the sciences legitimately accomplish and validate. It merely — but powerfully — complements our quest for understanding and explanation of origins by supplying a "barebones" but compelling resolution to the basic issue of the ultimate ground of existence and order.

Thus, quantum cosmological scenarios or theories — which describe the Planck era, and the Big Bang, or which describe the primordial regularities, processes, and transitions connected with these extreme very early stages of the universe — are in principle incapable of being alternatives to divine creation conceived as *creatio ex nihilo*. They simply do not account for what *creatio ex nihilo* provides — the ultimate ground of existence and order. Reciprocally, *creatio ex nihilo* is not an alternative to the processes and transitions quantum cosmology proposes and provides — these are models of the physical processes that generated our universe and everything emerging from it. *Creatio ex nihilo* by itself cannot, and was never intended to, usurp the role these, and the laws of nature upon which they depend, play in the universe. Instead, they are precisely the material, physical expressions, and channels of its continuing operation. Thus, quantum cosmology and *creatio ex nihilo* contribute deeply complementary and consonant levels of understanding of the reality in which we are immersed. Exactly the same point can be applied to divine creation and biological evolution — they are not exclusive alternatives, but rather complementary accounts, linking the ultimate ground of being and order with their elaboration in concrete structures, dynamisms, processes, and transitions.

181

Fr. William R. Stoeger, S.J. (USA), is an expert on the mathematical modeling of physical events associated with the Big Bang. He is also active in the fields of science and theology, and science and philosophy. This article is a highly abbreviated version of the chapter, "The Big Bang, Quantum Cosmology and Creatio ex Nihilo," to appear in Creation and the God of Abraham, *edited by Janet Soskice, David B. Burrell, Carlo Cogliati, and William R. Stoeger, forthcoming from Cambridge University Press.*

MATHEMATICS: THE LANGUAGE OF ASTRONOMY AND THE MIND OF GOD

• Fr. ANDREW P. WHITMAN, S.J. •

Above: *"Where were you when I laid the foundation of the earth? Tell Me, if you have understanding. Who determined its measurements — surely you know! Or who stretched the line upon it?" (Job 38:4-5)*
Credit: *Bible Moralisée, c. 1250. Osterreichische Nationalbibliothek, Wien.*

182

It is not too far from the truth to say that the beginnings of the Vatican Observatory lay with the promulgation of the Gregorian Calendar in 1582, the work of the famous Jesuit mathematician, Christoph Clavius, as described in Juan Casanovas' chapter of this book ("Astronomy, Calendars, and Religion"). Mathematics stills plays a critical role in astronomy. In particular, astrophysicists and cosmologists express their ideas on the origins and evolution of creation in terms of mathematics. Thus, it is not surprising that there is a strong mathematical component to the Vatican Observatory.

In fact, some 46 years ago, as one such professional mathematician on the staff of the observatory, I helped start (with Prof. Lawrence Conlon of Washington University, St. Louis) a community of mathematicians in service to the Church, which we baptized the "Clavius Group of Mathematicians." Every summer this group meets for four to six weeks at various locales: six times we have met at the *Institute des Hautes Études Scientifiques* outside of Paris; six times at the Institute of Advanced Studies in Princeton, New Jersey; four times at the *Centro de Investigación del Instituto Politécnico Nacional* in Mexico City; six times at Notre Dame University at South Bend, Indiana; five times at Boston College in Boston, Massachusetts; and in other years, at various other sites, usually Jesuit institutions. When the new quarters in the Vatican Gardens of Castel Gandolfo for the observatory are ready, we are hoping that the Clavius Group will hold their Summer Session in 2011 in these new quarters. In fact, as the many younger Jesuits who are preparing careers in astronomy, astrophysics, and cosmology are brought onto the active staff of the observatory, we can see an active collaboration between between them and the Clavius Group, broadening the vision of both.

But just what is this area of human knowledge called mathematics?

The simple process of counting gives us a clue. We can count: one cow, two cows, three cows, as well as one cat, two cats, three cats. But we quickly realize that we can count one, two, three without adding on the nouns "cow" or "cat." The world of mathematical abstraction seems immediate. But can we say more? Can we affirm more? Does this world of mathematics really depend on the physical world? Certainly we must affirm that it is not independent of the physical world, as the example above shows. However, the moment one gets into the more advanced structures of modern mathematics, it seems that these structures go beyond an abstraction from the physical world. Yet it never gets totally divorced from it, and in some manner is embodied in the physical world.

As an example, we suggest the following. Everyone in the science and engineering programs today has been challenged by a course in calculus. One of the two main pillars of that course is the Theory of Riemannian Integration of functions. (The other pillar is the Theory of Differentiation of functions.) Using much geometric intuition — and thus physical intuition — these theories were developed.

But it became evident in the latter half of the 19th century that these theories were inadequate, because certain very natural functions, when integrated, did not give the expected results. What was needed was another fundamental definition of integration. After much struggle and abstract reasoning — that is, without returning to the physical world for guidance — the method called Lebesgue Integration was discovered. (This fascinating struggle of moving from Riemann Integration to Lebesgue Integration can be read in the recent book of David Bressoud, *A Radical Approach to Lebesgue's Theory of Integration*, published by Cambridge University Press in 2008.)

Without going into the details of the mathematics itself, fascinating as they are, it is worth noting an essential facet of this development: the need for *abstract* reasoning, not tied to the physical world. It seems that only creatures with a capacity to reason abstractly can do mathematics. However, the basic question is still before us: where do these mathematical structures come from?

A suggestion to an answer to this question is found in a charming dialogue between the renowned neurobiologist Jean-Pierre Changeux and the mathematician Alain Connes. (It is found in a book by Changeux and Connes, *Conversations on Mind, Matter, and Mathematics*, translated by M. B. DeBevoise, published by Princeton University Press in 1995; see pages 260ff. We might mention that Connes is a winner of the Fields Medal in 1982, the equivalent of the Nobel Prize in Mathematics.) Both are materialists, and both seem to abhor any mention of divinity in explorations of the question: What is the nature of mathematics? But they mutually represent diametrically opposed positions in stating their answer to this question.

For Changeux, the neurobiologist, our brains have created all of mathematics. For Connes, mathematical reality and the harmony that characterizes it exist independently of the human brain, and actually preexist the material creation; it is independent of time and space. One can say that, philosophically, Connes is a Platonist.

The reductionist hypothesis of Changeux seems just impossible to many people. But if the only reality that one admits to is the created reality of our physical world, then of course one has no other resources for answering this question.

On the other hand, Connes is immediately put on the defensive, for if mathematics is not in our brain and is outside of creation and time and space, then he is faced with the question: Where is it? In fact, at first glance, his position seems to be absurd.

The Platonist assumption, however, fits nicely with the theist assertion that a God does really exist. In this conception, the world of mathematics is a reflection of the "Mind of God." It is totally immaterial — pure thought in the world of ideas. The fact that so much of this world of mathematics is applicable to the real physical world, such as the universe of the astronomers, is just an amazing confirmation of this world of mathematics.

But how much larger is this world than just those parts that we have seen as applicable to the real physical world? It is simply surprising that so much of the cosmology that Stoeger exposes in his essay in this collection ("Is Big Bang Cosmology in Conflict with Divine Creation?") must be communicated using mathematical language. And yet the part that he uses to express the cosmology is just a tiny fraction of the world of mathematical ideas that mathematicians explore.

Thus, one seems to demand that this world of mathematics is something objective, independent of the knowing mind. Indeed, it does fit the Platonic description of Connes. But to posit something like this begs for a source that sustains this reality. And the most satisfying hypothesis is that mathematicians are exploring the objectivity of a Creator God. How often has one heard over the centuries the exclamation that "God is a Mathematician!"

(If one is interested in exploring further this remarkable world of mathematics, it is recommended that one page through the book Mathematics: Frontiers and Perspectives, *edited by V. Arnold, M. Atiyah, P. Lax, and B. Mazur, and published in 2000 by the International Mathematical Union. I recommend skipping the more technical articles but reading some of the very profound philosophical ones.)* ●

Fr. Andrew P. Whitman, S.J. (USA), is a mathematician specializing in Lie algebra. He taught for many years with the Jesuits in Brazil before returning to the Vatican Observatory in Tucson.

THE POPES AND ASTRONOMY

A stronomy has long featured in Christian theology. Indeed, astronomy was one of the seven subjects of the medieval university that all scholars were expected to master before they could begin their studies of philosophy and theology. At the beginning of this book, we examined two specific instances in the history of the Church and astronomy: the successful reform of the calendar under Pope Gregory XIII in 1582, and the tragic conflict just 50 years later between the Church and Galileo. Here, however, we would like to take a look at more recent statements of Popes concerning the modern science of astronomy.

Much of the Church's interest has had an overt apologetic slant, using science to support its philosophical ideas or using its support of science to refute those who would accuse the Church of opposing progress and fearing newly discovered truths. Even in Roman times, the apologetic need for the Church's teachers to have an up-to-date knowledge of the physical universe, to give credibility to the theological truths of the Church, was evident to St. Augustine. Writing in A.D. 400, he commented:

Above: *By a happy coincidence, the coat of arms of Pope Leo XIII, the founder of the modern Vatican Observatory, included a comet! (It comes from his Pecci family coat of arms.)*
Opposite: *The Tower of the Winds during the papacy of Pius VII in the early 19th century. It housed an earlier version of the Specola Vaticana, set up under his predecessor Pius VI.*

Even a non-Christian knows something about the Earth, the heavens, and the other elements of this world, about the motion and orbit of the stars and even their size and relative positions, about the predictable eclipses of the Sun and Moon, the cycles of the years and the seasons ... and this

184

...that everyone might see clearly that the Church and her Pastors are not opposed to true and solid science ... but that they embrace it, encourage it, and promote it ...

Pope Leo XIII

The irony is, of course, that the cosmology the learned men of Rome knew so well was the very Ptolemaic cosmology later overthrown by Copernicus and Galileo!

But through the writings of these modern Popes, one begins to see developing a second realization: that, as the Psalmist knew, the heavens themselves do proclaim the greatness of the Creator. The simple act of seeking truth in the natural sciences is, in and of itself, a religious act, independent of any apologetic agenda.

From AETERNI PATRIS, 1879 (POPE LEO XIII)

In an encyclical letter proclaimed in 1879, subtitled "On the Restoration of Christian Philosophy in Catholic Schools in the Spirit (ad mentem) of the Angelic Doctor, St. Thomas Aquinas," Pope Leo XIII endorsed the study of Scholastic philosophy and ignited a new interest in the rational understanding of the faith. In passing, he reflects on the role of the physical sciences, in a way that foreshadows his establishment, 12 years later, of the Vatican Observatory itself:

Our philosophy can only by the grossest injustice be accused of being opposed to the advance and development of natural science. For, when the Scholastics, following the opinion of the holy Fathers, always held in anthropology that the human intelligence is only led to the knowledge of things without body and matter by things sensible, they well understood that nothing was of greater use to the philosopher than diligently to search into the mysteries of nature and to be earnest and constant in the study of physical things. And this they confirmed by their own example; for St. Thomas, Blessed Albertus Magnus, and other leaders of the Scholastics were

knowledge he holds to as being certain from reason and experience. Now, it is a disgraceful and dangerous thing for an infidel to hear a Christian, presumably giving the meaning of Holy Scripture, talking nonsense on these topics; and we should take all means to prevent such an embarrassing situation, in which people show up vast ignorance in a Christian and laugh it to scorn."

[St. Augustine, *The Literal Meaning of Genesis, Vol. 1* (trans. John Hammond Taylor, S.J.). New York: Paulist Press, 1982, pp. 42-43.]

never so wholly rapt in the study of philosophy as not to give large attention to the knowledge of natural things; and, indeed, the number of their sayings and writings on these subjects, which recent professors approve of and admit to harmonize with truth, is by no means small. Moreover, in this very age many illustrious professors of the physical sciences openly testify that between certain and accepted conclusions of modern physics and the philosophic principles of the schools there is no conflict worthy of the name.

THE REFOUNDATION AND RESTRUCTURING OF THE VATICAN OBSERVATORY, 1891 (POPE LEO XIII)

Here is the text of Leo XIII's **Motu Proprio,** *a personal decree that re-established the Vatican Observatory. In it, he explains the apologetic need for supporting a scientific institution at that time, and also outlines the previous history of Papal support for astronomy.*

So that they might display their disdain and hatred for the mystical Spouse of Christ, Who is the true light, those borne of darkness are accustomed to calumniate her to unlearned people and they call her the friend of obscurantism, one who nurtures igno-

rance, an enemy of science and of progress, all of these accusations being completely contrary to what in word and deed is essentially the case.

Right from its beginnings all that the Church has done and taught is an adequate refutation of these impudent and sinister lies. In fact, the Church, besides her knowledge of divine realities, in which she is the unique teacher, also nourishes and gives guidance in the practice of philosophy which is essential to understanding the scientific foundations of knowing — to make its principles clear, to suggest the criteria necessary for rigorous research and for a systematic presentation of the results, to investigate the soul's faculties, to study life and human behavior — and she does this so well that it would be difficult to add anything worth mentioning and it would be dangerous to dissociate oneself from her teachings.

Furthermore, it is to the great merit of the Church that the legal code has been completed and perfected, nor can we ever forget how much she has contributed through her doctrine, her example and her institutions to addressing the complex issues arising in the so-called social sciences and in economics.

In the meantime the Church has not neglected those disciplines which investigate nature and its forces. Schools and museums have been founded so that young scholars might have a better opportunity to deepen those studies. Among the Church's children and ministers there are some illustrious scientists whom the Church has honored and assisted as much as she could by encouraging them to apply themselves with complete dedication to such studies.

Among all of these studies astronomy holds a preeminent position. It proposes to investigate those inanimate creatures which more than all others proclaim the glory of God and which gave marvelous delight to the wisest

of beings, the one who exulted in his divinely inspired knowledge, especially of the yearly cycles and of the positions of the heavenly bodies (Wisdom 7:19).

The Church's pastors were motivated, among other considerations, to see to progress in this science and to support its followers by the possibility that it alone offered to establish with certainty those days on which the principal religious solemnities of the Christian mystery should be celebrated. So it was that the Fathers at Trent, well aware that the calendar reform done by Julius Caesar had not been perfect so that time calculations had changed, urgently requested that the Roman Pontiff would, after consulting experts in the field, prepare a new and more perfect reform of the calendar.

It is well known from historical documents how zealously and generously committed was Our Predecessor Gregory XIII in responding to this request. He saw to it that at the place judged to be best for an observatory within the confines of the existing Vatican buildings an observing tower was constructed and he equipped it with the best instruments of those days. It was here that he held the meetings of the experts he had selected for the reform of the calendar. This tower still exists today and it brings back the memory of its illustrious and generous founder. The meridian constructed by Ignazio Danti from Perugia is to be found there. Along the meridian line there is a round marble tablet whose lines are designed with such wisdom that when the Sun's rays fall on them it becomes obvious how necessary it was to reform the old calendar and how well the reform conformed to nature.

That tower, a splendid memory to a Pontiff who is to be much praised for his contribution to the progress of literary and scientific studies, was, toward the end of the last century after a long period of inactivity, restored to its original use as an astronomical observatory by the auspicious orders of Pius VI. Through the initiatives of a Roman Monsignor Filippo Gilii, other types of research were also undertaken on terrestrial magnetism, meteorology, and botany. But, after the death in 1821 of this very capable scientist, this monument to astronomical research went into neglect and was abandoned. Right after this Pius VII died and the energies of Leo XII were completely taken up with the reform of studies in the worldwide Church, a huge undertaking aimed at promoting all branches of learning. Such a reform, which had already been planned by his immediate and immortal predecessor, came by his efforts to a happy ending with the Apostolic Letter, *Quod divina sapientia*. In this letter he established certain rules with respect to astronomical observatories, the observations which were to be made regularly, the daily list of data to be made, and the information that was to be distributed internally concerning discoveries made by others.

The fact that the tower in the Vatican was no longer used as an observatory, after others in Rome had been equipped for that very purpose, came about because those who were competent to judge were of the opinion that the nearby buildings, and especially the dome which crowns the Vatican basilica, would have obstructed observations. And so it was deemed preferable to have observatories in other higher places where unobstructed observations could be carried out.

It then happened that, after those observing sites along with the whole city of Rome fell into the hands of others, we were given, on the occasion of our 51st anniversary as a priest, many excellent instruments for research in astronomy, meteorology, and Earth physics, as well as other gifts. It was the opinion of the experts that no place was better to house them than the Vatican tower, where, it seems, Gregory XIII had already in some way made preparations. After having evaluated this proposal and having examined the structure itself of the building, the history of its past glories, and the equipment already gathered there, as well as the opinions of persons renowned for their knowledge and judgment, we were persuaded to give orders that the observatory be restored and that it be equipped with all that would be required to carry out research not only in astronomy but also in Earth physics and in meteorology. As to the lack of an unobstructed view of the heavens in all directions from this Vatican tower, we saw fit to consider providing the nearby ancient and solid Leonine fortification where there is a quite high tower which, since it rises on the summit of the Vatican hill, provides for complete and perfect observation of the heavenly bodies. We, therefore, added this tower to the one of Gregory and we had installed there the large equatorial

Above: Pope Leo XIII.

telescope for photographing the stars.

To this purpose we chose conscientious men, prepared to do all that was necessary for such an undertaking, and we proposed to them a most competent scholar in astronomy and physics, Father Francesco Denza of the Clerks Regular of St. Paul, also called the Barnabites. Relying on their dedicated work, we agreed wholeheartedly that the Vatican Observatory be chosen to collaborate with other renowned astronomical institutes in the project to reproduce from photographic plates an accurate map of the whole sky.

Considering the fact that we wish this work of restoring the Specola to be a lasting one and not one that terminates after a short time, we have established bylaws for it with rules to be observed both for internal administration and for the services which others require of it. Furthermore, we have appointed a Board of carefully selected persons whose responsibility it is to govern the observatory and they have the highest authority after our own for all decisions respecting the internal administration.

And so with the present letter we confirm those bylaws and that Board and we also assign the various jobs and all that, with our order or consent, has been done with respect to the Specola. And we desire that the Specola be considered at the same level as the other Pontifical Institutes founded to promote the sciences. In order to provide in a more secure way for the stability of this work, we even designate a sum of money which should suffice to cover the expenses required to keep it operating and to maintain it. Nevertheless, we trust that such a work will find its justification and support in the favor and help of Almighty God more than

in what humans can do. In fact, in taking up this work we have become involved not only in helping to promote a very noble science which, more than any other human discipline, raises the spirit of mortals to the contemplation of heavenly events, but we have in the first place put before ourselves the plan which we have energetically and constantly sought to carry out right from the beginning of our Pontificate in talks, writings, and deeds whenever we were provided the opportunity. This plan is simply that everyone might see clearly that the Church and her Pastors are not opposed to true and solid science, whether human or divine, but that they embrace it, encourage it, and promote it with the fullest possible dedication.

We wish, therefore, that everything that has been established and announced in the present letter will remain into the future confirmed and ratified as it is proposed herein and we declare null and void any attempt at changes by whatsoever person. And it remains established and confirmed, despite any previous contrary declaration.

Given in Rome at St. Peter's,
March 14, 1891

DEUM CREATOREM VENITE ADOREMUS, 1935 (POPE PIUS XI)

As anyone with a smattering of Latin and Christmas carols will recognize, the title of this discourse can be translated as "Come, let us adore God the Creator." This text was delivered by Pius XI on September 29, 1935, at Castel Gandolfo on the occasion of the inauguration of the new Vatican Observatory headquarters.

We are very happy and very grateful to God that we can be present among you, my beloved

sons and daughters, to rejoice in the inauguration of this new, and might we say improved, Specola Vaticana in this our residence at Castel Gandolfo, it itself having been renovated.

It is not just for the sake of using the simple usual expression, but rather deliberately and with reflection, that we say: We are very happy and very grateful to God.

Today we officially inaugurate the Astronomical Observatory and

Cortie S.J.
Mons. Migone

the Astrophysical Institute. The high quality of the scientific instrumentation which has been acquired and the proven expertise of the scientists who will use it will undoubtedly make some important contributions to the progress in research of a science which, as a study of the heavens, can be said to be sovereign among the sciences. But this is not the only reflection which brings joy to our hearts today.

What we are doing today, and your presence my beloved sons and daughters, makes it all the more beautiful and solemn, adds some lines to the truly golden and most glorious pages which have already been written about the history of the Roman Pontificate. It carries us, like winged Pegasus, through the centuries in an immense and magnificent world of things, of ideas, of happenings.

Our dear and most capable Father Stein has given us a simple, but wise and most tasteful, account of that world. His brief history has for a moment thrown light upon and opened our eyes to the unfathomable depths of the heavens. We have been able to capture and enjoy at least a few notes from that immense hymn from on high where the heavens and the heavenly bodies sing the glory and reveal the power and the wisdom and the infinite beauty of the Creator.

And one might say that the Creator Himself, He Who at the end of His work of creation was pleased and proclaimed that all was good, is in a special way pleased with the magnificence of the heavens and the stars.

In fact, the divinely inspired Text emphatically and repeatedly calls upon the heavens and the stars to praise and bless the Lord (Psalm 148:3; Daniel 3:63 and elsewhere) and the Creator Himself gives it the name Beautiful Star (Apocalypse 22:16). That same sacred text finds one of the happiest expressions of divine wisdom when, in the presence of those infinite multitudes of heavenly bodies, which the new modern instrumentation magnifies and multiplies, it sees God numbering the multitude of stars and listens to Him calling each by name, a prerogative which God reserves to Himself (Genesis 15:5; Psalm 146:4). Again it is the divinely inspired Text which sees the uncreated Wisdom (Wisdom 7:29) shining sovereign among the stars: even more, in the beauty of the heavens and in the glory of the stars it sees God Himself Who illuminates the world from on high (Ecclesiastes 43:4). It is always the divine word which puts in the mouth of the disciple of Wisdom a special thanks for the knowledge that has been acquired of the stars (Wisdom 7:19).

We should not be amazed then if the magnificent matters which astronomy studies and helps us to better understand, and if the ideas which are raised by even the most ordinary but solid view of those matters, become the source of a profound spirituality. I am referring to the relationship between Religion and the Science of the heavenly bodies which has reigned continuously over the centuries from the most remote antiquity to our times. The most recent important Congress of the Orientalist in Rome has also reminded us of this relationship and treated of it in some of the papers. Information from ancient cuneiform and hieroglyphic texts have helped us to understand how observations of the heavens were related to sacrifices and cultic practices. All of this is now well known. It is only yesterday, in comparison with those antiquities, that the reform of the calendar took place. It is well known that the astronomy of those times played an important role in this reform which bears the name of one of our great predecessors — Gregory XIII. Even today that role is very much appreciated by such competent judges as a Schiaparelli and a Father Hagen, not to speak of others whom we know personally and admire.

J. Cerulli   Stein S.J.

en S.J.   Rodes S.J.

Above: Pope Pius XI with members of the Vatican Observatory on the occasion of the meeting of the International Astronomical Union in Rome in 1922. Immediately to his left is Fr. Johann Hagen, the director of the observatory.

From no part of Creation does there arise a more eloquent or stronger invitation to prayer and to adoration.
Pope Pius XI

That which we are doing now is more than all of that. We are taking up once more one of the threads of the history of the Roman Pontificate and it is a beautiful and precious one. That is the thread of its relationship over the centuries with the science of the heavenly bodies, a science one might in all truth say is by its nature religious, just like, as Tertullian so nicely phrased it, the human soul is naturally Christian. In fact, from no part of Creation does

ing as the immense vault of the sky passes over. Even the unbelieving poet [Carducci] in the silence of the starry heavens was want to hear passing over the soft sweet prayer of the Ave Maria. It seems to us, dear sons and daughters, in this astronomical, may we call it, inauguration that we are fulfilling in the name of the whole Church, an act of our priestly ministry.

With a very fortunate thought the author of the new Specola, Father Stein,

It is also quite well known that the Supreme Roman Pontiffs have for many centuries needed astronomy and have called upon it to help in the placement of holy temples and especially in the calculation of the date of Easter.

As you see, what we are doing here is not just to continue and to imitate, within our resources, the patronage of our illustrious predecessors who have never been sufficiently praised for what they did. It is not just that we are trying to assure for the present and future, as they did for the past with the quiet eloquence of their accomplishments, to assure I repeat that implicit, even explicit, defense of the Faith and of Religion. That defense shines and is more than ever persuasive whenever respect for the faith is joined in a spontaneous way with the development of Science.

there arise a more eloquent or stronger invitation to prayer and to adoration. As the Wise Men of old, to whom the stars announced the coming of God to the earth, expressed it: We have seen His star and have come to worship Him. Even today the Beduin of the vast deserts sees the majesty of God shin-

Above: Pope Pius XI at the dedication of the new telescopes on the roof of the Papal summer residence in Castel Gandolfo, 1935.

recalled for us the imposing inscription ordered by Pius IX to be placed on the Pontifical Observatory of the Roman University on the Capitoline, an observatory which he had constructed: *Deo Creatori* (to God, the Creator). We could do no better than to follow in the wake which our glorious predecessor so brilliantly left open and to complete his thought by ourselves declaring and inscribing on the new Specola: *Deum Creatorem Venite Adoremus* (Come, let us adore God the Creator).

From AN ADDRESS
to the PONTIFICAL ACADEMY
OF SCIENCES, 1939
(POPE PIUS XII)

The Pontifical Academy of Sciences can be traced back to the Academy of the Lynxes, founded in 1603 by Prince Federico Cesi under the patronage of Pope Clement VIII; it claimed Galileo as one of its most notable members. In 1923, Pope Pius XI gave the academy its current building within the Vatican Gardens, and in 1936 he reorganized the academy as a "Scientific Senate" composed of leading scientists of the world, invited from all nations and faiths, chosen "without racial or religious discrimination." One of the churchmen he appointed to this academy was Eugenio Cardinal Pacelli, who would later become his successor, Pope Pius XII.

Soon after his election as Pope, in December 1939 Pope Pius XII addressed the Plenary Session of the academy. In it, he endorsed the work of the academy, calling it his predecessor's "greatest achievement." In these excerpts from his opening address, he begins by quoting from the same book of Augustine on Genesis that we referenced above:

As St. Augustine tells us, God, having created the universe, did not abandon the world but kept man's thoughts in His counsel. While maintaining the universe in existence and motion, God left it to men to dispute amongst themselves without their being able to discern God's full project. God has given fallen man this task of understanding this great enigma; the enigma of the unknown God working in creation, to which Paul the Apostle pointed when addressing the Epicurean and Stoic philosophers in the Athenian council of the Areopagus. Paul stated that this unknown God had created the whole human race on the entire Earth so that they could find their way towards God since He is not

Top: Pope Pius XI at the dedication of the new spectrochemical laboratory on the ground floor of the Papal summer residence in Castel Gandolfo, 1935. Third from the left in the group of priests to the Pope's left is Fr. Ledókowski, the Superior General of the Jesuits.

far from any of us.

The enigma of creation has for centuries stretched the intellect of all peoples; the various solutions proffered have filled the schools of the academy; volumes have filled both ancient and modern libraries; attempts to find the solution to this enigma have been the cause of disputes between wise investigators of nature, of matter, and of the spirit. These labors, these lessons, these volumes, these battles are nothing other than the searchings for the truth hidden deep in the enigma itself....

Reality speaks to us and communicates her word to us through the wonderful sense of our nature moulded out of flesh and spirit. It is this reality which we seek through the immeasurable ways of the universe. We are neither responsible for creation nor are we the creators of Truth; neither our doubts, nor our opinions, nor our carelessness, nor our negations can alter it.... Our human investigations measure the truth found by our scientific implements and instruments and various machines; they are able to transform, capture and dominate the material offered to us by nature, but they cannot create her; our minds have to remain faithful in following nature just as a disciple does with his master from whom he learns his work...

Man ascends to God by climbing the ladder of the Universe: the astronomer, when reaching the sky, footstool to the throne of God, cannot remain an unbeliever before the voice of the firmament; from beyond the suns and astral nebulae emanates the thought, followed by the love and adoration, which sails

> *Man ascends to God by climbing the ladder of the Universe.*
> **Pope Pius XII**

and unveiling the secrets of nature and teaching men to direct their energies for their good, at the same time you preach, in the language of numbers, formulae, and discoveries, the ineffable harmonies of the all-wise God.

Contrary to rash statements in the past, the more true science advances, the more it discovers God, almost as though He were standing, vigilant and waiting, behind every door which science opens. Furthermore, we wish to

> *The more true science advances, the more it discovers God, almost as though He were standing, vigilant and waiting, behind every door which science opens.*
> **Pope Pius XII**

toward a sun which illuminates and gives warmth not to the clay of man but to the spirit which animates him.

From **THE PROOFS FOR THE EXISTENCE OF GOD IN THE LIGHT OF MODERN NATURAL SCIENCE, 1951 (POPE PIUS XII)**

*I*s it possible for the Church to endorse science too much? That became an issue following the comments of Pope Pius XII given below. Speaking to the Plenary Session and to the Study Week on the subject "The Question of Microseisms," the Pope addressed at length the structure of matter and the cosmos and the origins of the universe, using them to re-examine classical proofs of the existence of God on the basis of new scientific discoveries. We include here his opening comments and the section in which he dealt with modern astronomy:

We are grateful to the Almighty for a serene hour of happiness which offers us this gathering of the Pontifical Academy of Sciences, and gives us the welcome opportunity of meeting with a select group of Eminent Cardinals, of illustrious diplomats and of noteworthy personalities, and especially with you, Pontifical Academicians, who are truly worthy of the solemnity of this session; because in investigating

say that not only does the philosophical thinker benefit from this progressive discovery of God, achieved in the increase of knowledge — and how could he do otherwise? — but those also profit who participate in the new discoveries or who make them the object of their considerations. The genuine philosophers

especially benefit from it, since, by using the scientific advances as a springboard for their rational speculations, they can achieve greater security in their conclusions, clearer illustrations in possible obscurity, more convincing support in finding ever more satisfactory answers to difficulties and objections....

The Universe and Its Development

In the future: If, then, the scientist turns his gaze from the present state of the universe to the future, however far off, he will be forced to realize that the world is growing old, both in the macrocosm and in the microcosm. In the course of billions of years, even the quantity of atomic nuclei, which is apparently inexhaustible, loses its utilizable energy and matter approaches, to speak figuratively, the state of a spent and wasted volcano. And the thought presents itself inescapably: if the present cosmos, today so pulsating with rhythm and life, is not sufficient to account for its existence, as we have seen, how much less will it be the case for that cosmos once the shadow of death shall have passed over it.

In the past: We now turn our eyes toward the past. In proportion to the distance in time to which we turn backward, matter is seen to be richer and richer in free energy and the theater of great cosmic upheavals. Thus, everything seems to indicate that the material universe has had, in finite time, a powerful start, provided as it was with an unimaginable abundance of reserves in energy; then, with increasing slowness, it has evolved to its present state.

Two questions spontaneously come to mind:

Is science in a position to say when this powerful beginning of the cosmos took place? And what was the initial, primitive state of the universe?

The most noted experts in atomic physics, in cooperation with the astronomers and the astrophysicists, have put great effort into shedding light on these two difficult but extremely interesting problems.

The Beginning in Time

First, to cite some figures, which serve only to express the order of magnitude in the designation of the dawn of our universe, that is, its beginning in time, science has at its disposal several paths of investigation, each fairly independent of the other, though they are convergent, as we indicate briefly:

1. The velocity of travel of the spiral nebulae or galaxies: The examination of numerous spiral nebulae, carried out especially by Edwin E. Hubble at Mt. Wilson Observatory, has demonstrated the significant result — though tempered by reserve — that these far-off systems of galaxies tend to rush away from one another at such speed that the space between two such spiral nebulae doubles in the period of about 1,300 million years. If one looks back across the period of this process of the "Expanding Universe" the conclusion is that from one to ten billion years ago the matter of all the spiral nebulae was compressed into a relatively narrow space, at the time of the beginning of the cosmic processes.

2. The age of the solid crust of the Earth: To calculate the age of the original radioactive substances, highly approximate data are deduced from the transmutation of these substances into the corresponding isotope of lead, for instance the transformation of the isotope of uranium 238 into RaG (an isotope of lead), of the uranium isotope 235 into actinium D, and of the isotope of thorium 232 into thorium D. The mass of helium which is formed thereby can also serve as a check. In this way the average age of the most ancient minerals is indicated at a maximum of five billion years.

3. The age of meteorites: The preceding method, when applied to meteorites to calculate their age, gives about the same figure of five billion years. This result takes on special importance because the meteorites are generally believed to be of interstellar origin and, except for terrestrial minerals, they are the only examples of celestial bodies which can be studied in scientific laboratories.

4. The stability of the systems of double stars and star masses: The oscillations of gravitation within these systems, like the wearing away of the tides, again restrict their stability within the limits of from five to ten billion years.

Although these figures are astonishing, nevertheless, even the simplest believer would not take them as unheard

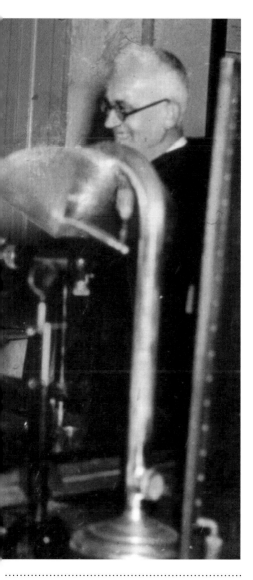

Above: Pope Pius XII visiting the Vatican Observatory Spectrochemical Lab in 1948. With him are the Vatican astronomers Fr. Gatterer (left) and Fr. Junkes (right).

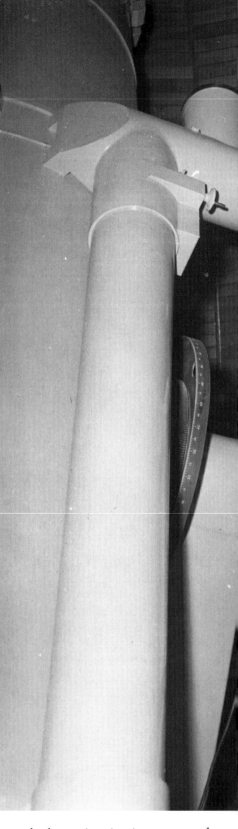

degrees of enormous intensity, as can be seen in the recent work of A. Unsold, director of the observatory in Kiel. Only under these conditions can one comprehend the formation of the heavy nuclei and their relative frequency in the periodical system of the elements.

On the other hand, the eager mind, in its search for truth, rightfully insists upon asking how matter came to be in a state so unlike that of our common experience of today, and what preceded it. One waits in vain for an answer from natural science, which honestly declares that this is an insoluble enigma. It is true that this is asking too much of natural science as such; but it is also true that the human spirit versed in philosophical speculation is able to penetrate the problem more profoundly.

It is undeniable that a mind illuminated and enriched by modern scientific knowledge, which calmly evaluates this problem, is led to break the circle of a matter preconceived as completely independent and autonomous — either because uncreated or self-created — and to acknowledge a Creative Spirit. With the same clear and critical gaze with which he examines and judges facts, he also catches sight of and recognizes the work of the omnipotent Creator, whose power, aroused by the mighty "fiat" pronounced billions of years ago by the Creative Spirit, unfolded itself in the universe and, with a gesture of generous love, called into existence matter, fraught with energy. Indeed, it seems that the science of today, by going back in one leap millions of centuries, has succeeded in being a witness to that primordial Fiat Lux, when, out of nothing, there burst forth with matter a sea of light and radiation, while the particles of chemical elements split and reunited in millions of galaxies.

It is true that the facts verified up to now are not arguments of absolute

of and differing from those derived from the first words of Genesis, "In the beginning ...," which signify the beginning of things in time. These words take on a concrete and almost mathematical expression, and new comfort is given to those who share with the Apostle an esteem for that Scripture, divinely inspired, which is always useful: *ad docendum, ad arguendum, ad corripiendum, ad erudiendum* — to teach, to prove, to correct, to educate.

The State and Nature of Original Matter

With equal earnestness and freedom of investigation and verification, learned men, in addition to the question of the age of the cosmos, have applied their audacious talents to another question which we have already mentioned and which is certainly much more difficult, and that is the problem concerning the state and quality of primitive matter. According to the theories which are taken as a basis, the relative calculations differ considerably one from the other. Nevertheless, the scientists agree in holding that not only the mass but also the density, the pressure, and the temperature must have attained

> *Creation, therefore, in time, and therefore, a Creator....*
> **Pope Pius XII**

proof of creation in time as are those which are drawn from metaphysics and revelation, in so far as they concern creation in its widest sense, and from revelation alone in so far as they concern creation in time. The facts pertinent to natural sciences, to which we have referred, still wait for further

Above: Pope Pius XII at the finderscope of the Schmidt telescope, newly installed in the Papal Gardens of Castel Gandolfo, in 1957. He held a private ceremony for the astronomers of the Vatican Observatory, blessing their instrument and their work.

investigation and confirmation, and theories founded upon them have need of new developments and proofs, in order to offer a secure basis to a line of reasoning which is, of itself, outside the sphere of the natural sciences.

Notwithstanding this, it is worth noting that modern exponents of the natural sciences consider the idea of the creation of the universe entirely reconcilable with their scientific conception, and indeed they are spontaneously brought to it by their researches, though only a few decades ago such a "hypothesis" was rejected as absolutely irreconcilable with the present status of science. As late as 1911, the celebrated physicist Svante Arrhenius declared that "the opinion that something can proceed from nothing is in contrast with the present status of science, according to which matter is immutable." Similar to this is Plato's affirmation: "Matter exists. Nothing proceeds from nothing: in consequence matter is eternal. We cannot admit the creation of matter."

How different and reflecting great vision is the language of a modern top-grade scientist, Sir Edmund Whittaker, a Pontifical Academician, when he speaks of his researches concerning the age of the world: "These different estimates converge to the conclusion that there was an epoch about 10^9 or 10^{10} years ago, on the further side of which the cosmos, if it existed at all, existed in some form totally unlike anything known to us: so that it represents the ultimate limit of science. We may perhaps without impropriety refer to it as the Creation. It supplies a concordant background to the view of the world which is suggested by the geological evidence, that every organism ever existent on the Earth has had a beginning in time. If this result should be confirmed by later researches, it may well come to be regarded as the most momentous discovery of the age; for it represents a fundamental change in the scientific conception of the universe, such as was effected four centuries ago by the work of Copernicus."

Conclusion

What, then, is the importance of modern science in the argument for the existence of God drawn from the mutability of the cosmos? By means of exact and detailed investigations into the macrocosm and the microcosm, it has widened and deepened to a considerable extent the empirical foundation upon which the argument is based and from which we conclude a self-existent Being (*Esse per essentiam*) immutable by nature. Further, it has followed the course and the direction of cosmic developments, and just as it has envisioned the fatal termination, so it has indicated their beginning in time at a period about five billion years ago, confirming with the concreteness of physical proofs the contingency of the universe and the wellfounded deduction that about that time the cosmos issued from the hand of the Creator.

Creation, therefore, in time, and therefore, a Creator; and consequently, God! This is the statement, even though not explicit or complete, that we demand of science, and that the present generation of man expects from it. It is a statement which rises from the mature and calm consideration of a single aspect of the universe, that is, of its mutability; but it is sufficient because all mankind, the apex and rational expression of the macrocosm and the microcosm, is made conscious of its sublime Creator and feels His presence in space and in time, and, falling to its knees before His sovereign Majesty, begins to call upon the name *Rerum Deus, tenax vigor — Immotus in te permanens — lucis diurnae tempora — successibus determinans.*

The knowledge of God as unique Creator, a conviction shared by many

The human spirit has surpassed all the limits of the body's senses ... and succeeded in seizing the immense Universe.
Pope Pius XII

The divine Spirit reveals itself from the coldness of space to the scientist open to finding a purpose for the whole of existing reality.
Pope Pius XII

This conviction, which takes into account the deepest movements of science, is crowned by faith which, the more it is rooted in the consciousness of peoples, the more it can really lead to a fundamental progress for civilization.

It is a whole vision, of the present and of the future, of matter and of spirit, of time and of eternity, that, illuminating the mind, will save the men of today from a long and stormy night. And that faith, which makes us in this moment raise to Him Whom we have just called Vigor, Immotus and Pater, a fervent prayer for all His sons, who are given to us to look after:

Largire lumen vespere — quo vita nusquam decidat — light for our life in time, light for eternal life.

modern scientists, is certainly the extreme limit which natural reason is capable of reaching; but it does not constitute the last frontier of truth. Science, which has encountered the Creator in its path, philosophy, and, much more, revelation, in harmonious collaboration because all three are instruments of truth, like rays of the same sun, contemplate the substance, reveal the outlines, and portray the lineaments of the same Creator. Revelation especially renders the presence almost immediate, full of life and love, which is what the simple believer and the scientist are aware of in the intimacy of their spirits when they repeat without hesitation the concise words of the ancient Creed of the Apostles: *Credo in Deum, Patrem omnipotentem, Creatorem caeli et terrae!*

Today, after so many centuries of civilization (because they were centuries of religion), now the need is not to find God for the first time, but rather to recognize Him as a Father, to revere Him as Legislator, to fear Him as Judge; it is urgent for the salvation of all peoples that they adore the Son, the loving Redeemer of mankind and they bend the knee to the gentle urgings of the Spirit, fruitful Sanctifier of souls.

The Pope's text is careful to state that "the facts verified up to now are not arguments of absolute proof of creation in time ... [but] still wait for further investigation and confirmation, and theories founded upon them have need of new developments and proofs, in order to offer a secure basis to a line of reasoning which is, of itself, outside the sphere of the natural sciences." Nonetheless, it appeared to many to be an endorsement of a particular cosmology theory, popularly known as the "Big Bang."

The most startling reaction against this speech came from Fr. Georges Lemaître, the Belgian diocesan priest and astrophysicist whose work in the 1920s actually laid the foundation for the Big Bang theory. Lemaître was very leery at reading such an interpretation into his theory.

For one thing, he recognized that his theory was still controversial, and only one of several possible cosmologies, given the evidence available at that time. (It would be more than a decade before the discovery of the cosmic microwave background radiation finally convinced most cosmologists that some sort of "big bang" actually did occur.) Worse, in 1951, there was still some suspicion that this "priest's theory" might have been invented precisely to find a scientific excuse to support the Genesis description of "fiat lux." Lemaître wanted his work to be judged purely on its scientific merits. He knew that even the best scientific theory is eventually superseded by later work. And, in any event, while it is good that theologians be aware of the latest advances in science, it is certainly not theology's role to judge or endorse scientific theories. Nor, for that matter, is it wise to base theology on the latest advances of science, since that is a ground that is forever shifting.

ADDRESS TO THE GENERAL ASSEMBLY OF THE INTERNATIONAL ASTRONOMICAL UNION, 1952 (POPE PIUS XII)

Lemaître spoke personally to the Pope about his concerns, and clearly he was heard. The following year, when the International Astronomical Union (of which the

Vatican is a member state) met in General Assembly in Rome, the Pope was invited to address the assembled astronomers; in his address, presented on September 8, 1952, no such endorsement of the Big Bang or any other particular theory was put forward.

Instead, the Pope presented an up-to-date overview of the state of astronomical knowledge of the time, prepared with the assistance of the director of the Vatican Observatory, Fr. Daniel O'Connell, S.J. And then he reflected on the deeper implications of the very nature of the astronomical enterprise.

This hauntingly beautiful appreciation of the love of the Universe marked an important transition in the nature of the Church's appreciation of astronomy. No longer was astronomy merely an apologetic tool for the Church; now it was recognized to be in itself a profound act of worship of the Creator.

Here is a modern translation from the French text, prepared by the Jesuit astronomer Fr. Paul Gabor, S.J., who has included a note (appended as an italicized passage at the end) to describe how modern astronomy has developed beyond the situation cited by Pope Pius XII in 1952.

I. The Cosmic Panorama

The presence of such a myriad gathering of renowned astronomers from all over the world brings to our mind an image of the panorama of the Universe obtained by modern astronomy, which you have brought to its current perfection thanks to your untiring observations and brilliant analysis. We are grateful to you for many reasons, but in particular because your scientific research of the Universe and its exalting contemplation brings our spirits to philosophical considerations of a more universal value and lifts them up ever more to the knowledge of the End in its ultimate truth which surpasses all knowing and marks all existence with

its seal: "The Love which moves the Sun and other stars" (Dante, *Paradiso*, 33, 145).

Although we are aware that we speak before elite representatives of science, much more knowledgeable on the subject than we are, we cannot refrain from recalling at least in broad lines the admirable progress of astronomy and astrophysics over the last fifty years, and to indicate the milestones which will thus serve as a basis for our higher considerations.

What once was an enigma and a dream for the astronomers of the past, and what has become for the astronomers of the present age a reality beyond all our dreams, may perhaps be expressed with some justice as the conquest of space. Observations, understanding, and new technological means have, as it were, put a gigantic compass into the hands of astronomical science, opening up the Universe more every day, to the point where it now can embrace dimensions surpassing all expectation. How many barriers, especially those of enormous distances, have fallen over the last decades under the relentless pressure of the enquiring spirit, never satisfied: the spirit of the scientist!

The previous century witnessed the first laborious attempts at exploring the depths of space when Bessel, Struve, and Henderson measured the first trigonometric parallaxes; by the turn of the century one could feel a legitimate satisfaction at being able to determine with some certitude the distances of 58 fixed stars ranging up to 30 or 40 light-years from our Sun. After 1912, however, a new, different, and more

efficient method for measuring cosmic distances took human sight even further. Miss Leavitt discovered in a certain type of variable star, the Cepheids, a relationship between the period of their variability and their magnitude or brightness. Thus, wherever a Cepheid was discovered in the sky, one could deduce its absolute luminosity from the period of its variability, and comparing the absolute luminosity to the apparent brightness, easily calculate its distance.

At the same time, observations were given a boost by the growing sensitivity of photographic emulsions and by progress in the building of ever more powerful telescopes, which allowed the radius to which the human eye could penetrate to increase several million times, reaching unsuspected depths in space. The astronomer Shapley took the first great step beyond the closest stars with his classic research on the spatial distribution of globular clusters. This research led to a complete transformation of the idea of the galactic system's structure.

Meanwhile other inquiries, such as those dealing with stellar motions and the decrease of light when it passes through opaque matter in interstellar space, improved this new idea. We have thus acquired a certainty that the Milky Way of the ancients, which inspired so many naïve myths, is an immense accumulation of about a hundred thousand million stars — some larger, some smaller than our Sun — spanned by vast clouds of cosmic gas and dust. The whole system, obeying the general law of gravity, keeps revolving on gigantic orbits around a center located in the great stellar clouds of Sagittarius.

> *What a happy and sublime encounter over the contemplation of the cosmos is that of the human spirit with the Spirit of the Creator!*
> **Pope Pius XII**

Resembling in its whole an enormous spinning biconvex lens, this system has a diameter of about 100,000 light-years and a thickness of about 10,000 light-years at its center. As for us with our Solar System, we are not, as was believed earlier, at the center of this incommensurable accumulation of stars: in fact, we are about 30,000 light-years from it. And even though we revolve around it at the dizzying speed of about 250 km per second, it takes us 225 million of our solar years to make one whole revolution!

It is with legitimate pride that our century's astronomical science can claim to have conquered the galactic system. This first and happy leap forward was soon followed by another leap, which was to take human knowledge beyond the Milky Way into the immensity of space. It is mainly thanks to the gigantic telescopes of Lick, Yerkes and Mt. Wilson that this decisive step could be accomplished. When in 1917-19 Ritchey discovered several novae in the Andromeda nebula, few scientists believed the hypothesis that we

were dealing with stars in an extragalactic nebula at a distance of hundreds of thousands of light-years. It was only when Hubble, using the great two-and-a-half-meter mirror of Mt. Wilson, succeeded in resolving the external parts of the Andromeda nebula into isolated stars and globular clusters and in identifying several Cepheids that the adversaries surrendered their opposition. We have become certain that these spiral nebulae are in fact great stellar systems resembling our galactic system in their composition and their size, but so distant that they appear to the eye as mere tiny speckles of luminous haze. The distance of the nebula closest to us, Andromeda, was found to be 750,000 light-years; that of the Triangulum nebula, about 780,000. Tirelessly searching the skies, astronomers then started to look at other nebulae apparently much smaller than these galaxies and to calculate their respective distances, measuring their apparent diameters and brightnesses, and comparing these data with the known characteristics of the closest nebulae. Finally, Humason's spectroscopic research led to the discovery of an unsuspected law: the shift of spectral lines towards the red increases proportionately with a nebula's dis-

> *Science, which has encountered the Creator in its path, philosophy, and, much more, revelation, in harmonious collaboration because all three are instruments of truth, like rays of the same sun, contemplate the substance, reveal the outlines, and portray the lineaments of the same Creator.*
> **Pope Pius XII**

> *You have two works to do. One is to explain the world of science to the Catholic Church. The other is to explain the Catholic Church to your fellow scientists. I think you do the second much better!*
> **Informal comments of Pope John XXIII to Fr. Martin McCarthy, S.J., at the Vatican Observatory**

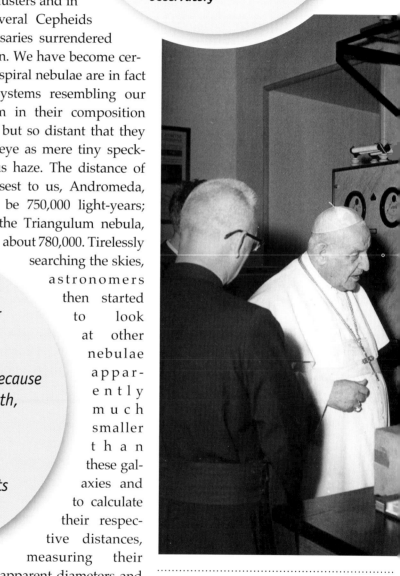

Above: Though giving no formal addresses concerning the world of astronomy, Pope John XXIII had a great love of the Vatican Observatory and the Jesuits who shared his summer home. In 1959, on the feast of St. Ignatius, founder of the Jesuits, he came to see the Astrophysical Laboratory (above) and the meteorite collection (right). Over the years, he paid other occasional

tance, so that the measure of this shift allows us to evaluate distance itself, despite the faintness of the light which reaches us, so long as it is sufficient to produce a measurable spectrum.

During this research, it was observed that — when considering broad and deep zones of the sky — these extragalactic nebulae appear as roughly equally distributed in cosmic space, and so far it has been impossible to observe the slightest decrease in their

informal visits to the community during his summer stays. The diary of Fr. O'Connell, then director of the observatory, notes two occasions when the Pope left with the community "the gift of an excellent wine."

density. In the space reached by the telescope on Mt. Wilson, we estimate the number of galaxies at about 100 million, distributed in a sphere of an approximate diameter of a thousand million light-years, each of them containing about a hundred thousand million stars similar to our Sun.

After this brief imaginary trip through the immensity of the Universe, let us return to our little planet which, with the mass of its mountain ranges, with the limitless span of its oceans and deserts, with the violence of its hurricanes, its volcanic eruptions and seismic movements, sometimes appears to us so vast and powerful. In fact, in the space of a second, a beam of light would circle our equator more than seven times. In a little more than a second — a blink of an eye — it would reach our neighbor, the Moon. In a little more than eight minutes, it would reach the Sun. And in five and a half hours, it would touch the farthest planet of our system, Pluto. As for the closest fixed stars, which during serene nights on mountaintops seem so close that we can almost touch them, a light message would take more than four years to reach them and it would need 30,000 years to reach the center of the Milky Way. The light coming to us from the Andromeda nebula

Lord, I am ready to walk in the night, step by step, enlightened by the stars.
Pope John XXIII

left its source about 750,000 years ago, whereas for certain very distant nebulae which only the most powerful modern optical instruments can detect with great difficulty on a photographic plate after a very long exposure, as tiny stars, are at a distance of 500 to 1,000 million light-years.

What numbers, what dimensions, what distances in space and time! And yet we have to believe that astronomical science is far from being considered as having reached the end of its marvelous adventure.

Who can say what future steps forward can be achieved in the near future by the five-meter mirror on Mt. Palomar, and the rapid development of radio astronomy? How small does man appear in the prodigiously extended framework of space and time: a minute speck of dust in the immensity of the Universe. And yet!

II. The Work of Inquiring Spirit

When we look at the picture of the Universe sketched above, a picture which is the fruit of long and laborious research by generations of researchers hailing from the most diverse nations, what is most striking beyond what has already been said is not so much the gigantic mass of the Universe and all its parts, or the harmony of its movement;

rather, it is the behavior of the inquisitive human spirit in discovering such a vast panorama. By its nature tied to the body's condition of minute dimensions, the human spirit has surpassed all the limits that the feeble power of the body's senses seemed to be capable of, and succeeded in seizing the immense Universe.

This is truly an enormous accomplishment. Consider the starting point of this admirable trek through the heavens. Our senses, with which this trek begins, are severely restricted by being generally limited to the sphere of space and time that immediately surrounds us. The first grace of the spirit was thus to overcome the narrow enclo-

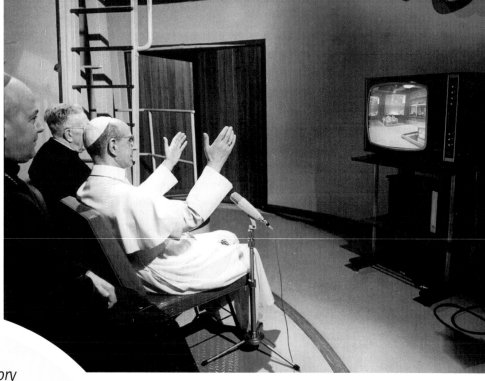

Here, from his observatory at Castel Gandolfo, Pope Paul VI is speaking to you astronauts. Honor, greetings, and blessings to you, conquerors of the Moon, pale lamp of our nights and our dreams! Bring to her, with your living presence, the voice of the spirit, a hymn to God our Creator and our Father. We are close to you, with our good wishes and with our prayers. Together with the whole Catholic Church, Pope Paul VI greets you.

Pope Paul VI to the Apollo astronauts, July 21 (Rome time), 1969

sure imposed on the senses due to the conditions of their own nature, by inventing the means of building ingenious instruments to increase the size and precision of their perception beyond all limits: the telescope, which nearly annuls the enormous distances between the eye and the far stars, making them present and as if within reach; and the photographic plate, which collects and fixes the faintest light of the farthest nebulae.

With this increase of the senses' power by the human mind, gradually the spirit has employed this gained power to deepen our research into nature, inventing a thousand ingenious methods to unveil the most subtle and hidden phenomena. Thus, for example, it sums up the smallest effects over and over in order to obtain a perceptible integrated signal, inventing instruments such as the photoelectric cell or the Wilson chamber to explore the finest atomic processes of radioactive matter and cosmic radiation. Searching ever more, it discovers the laws governing energetic processes and so can change forms of energy which are beyond the sphere of sensible perception — such as the electric waves of infra-red and ultra-violet radiation — into others which can enter into the realm of direct and very precise perception by the senses. The spirit puts nature to the test in laboratory experiments from which it deduces laws, valid provisionally over the restricted conditions of its attempts. Not satisfied with this, it experiments and extends the radius of these experiments' application to the realm of observational astrophysics. The practical and theoretical knowledge of molecular spectra enable it to reach into the dense atmospheres of the gas giant planets and to verify the composition, temperature, and density astronomical spectral lines even before it is possible to obtain them in the laboratory, and explains where they belong and where they come from. The depths of the Sun's sphere itself cannot escape its penetrating glance, armed with astrophysical theories. It follows the dissociation of matter there, the nuclear processes that take place in the center of the Sun and which serve to provide the energy emitted by its radiation over thousands of millions

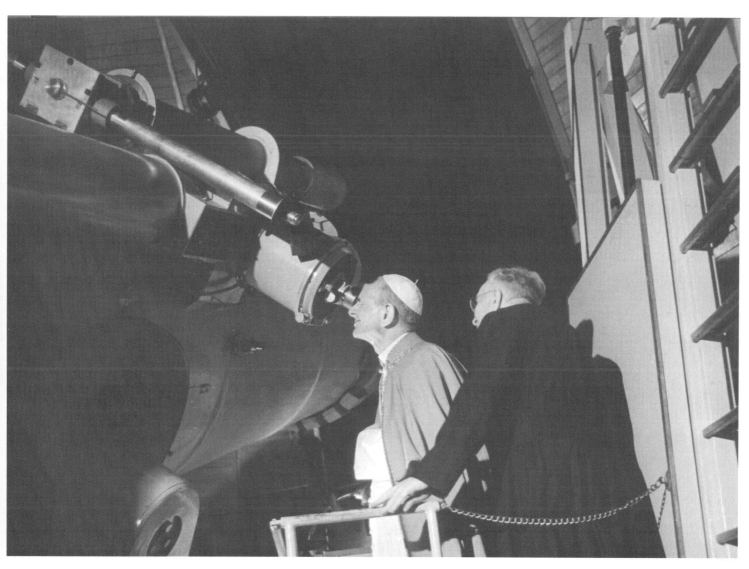

Left, above and below: Pope Paul VI addressing the Apollo astronauts on the Moon, as a part of a worldwide broadcast of world leaders to the Moon on July 21, 1969. He is speaking from the dome of the Vatican Observatory's Schmidt telescope.
***Right:** Pope Paul VI with the director of the Vatican Observatory, Fr. Daniel O'Connell, observing the Moon in the finderscope of the Schmidt telescope, the night of the first Moon landings.*

of their gas. Using the facts and the theories of spectroscopic science, it raises its inquisitive eyes to the fixed stars to reap exact knowledge of the composition, temperature, density, and ionization of their mysterious atmospheres. With the help of modern quantum theory, the investigative spirit interprets of years. The daring and brave human spirit cannot be stopped by the most formidable cataclysms of novae or supernovae. It measures the enormous speeds of ejected gas and tries to discover their causes. It embarks upon the trail of galaxies fleeing through space, retracing the journey they followed

> So
> much of our world
> seems to be in fragments....
> But at the same time
> we see a growing critical openness
> towards people of different
> cultures and backgrounds ...
> discovering values and experiences
> they have in common
> even within their diversities.
>
> **Pope John Paul II**

over thousands of millions of years of past time, and thus becomes akin to a spectator of cosmic processes that took place at the dawn of creation.

What is, then, the spirit of this tiny being which is man, lost in the ocean of the material Universe, that it dares to ask its infinitesimally small senses to discover the face and history of the immense cosmos, and to have uncovered them both? There is only one possible answer, only one overwhelming fact — the human spirit belongs to a category of Being fundamentally different from matter, and superior to it, even though matter may be of unlimited dimensions.

III. Eternal Creating Spirit

Finally a question comes spontaneously to mind: would the path to which the human spirit has committed itself, in a way which unquestionably honors it up to now, be always open before it? Will it follow this path without interruption until it discovers the last of the enigmas that the Universe keeps in store? Or, on the contrary, is the mystery of nature so ample and so hidden that the human spirit, because of its smallness and intrinsic disproportion, could never succeed in probing it completely? The answer of those powerful minds who have penetrated the deepest into the secrets of cosmos is quite modest and reserved: they think we are at the beginning. So much of the path remains to be followed. It will be followed unremittingly; nonetheless, there is no likelihood that even the greatest researcher could ever succeed in knowing, and even less in solving, all the mysteries contained within the physical Universe. One may therefore postulate the existence of a Spirit infinitely greater, a divine Spirit that creates, conserves, governs, and thus knows and penetrates in one supreme intuition, now as at the dawn of the first day of creation, all that exists: *Spiritus Dei ferabatur super aquas* (the Spirit of God moved upon the face of the waters) (Genesis 1:2).

What a happy and sublime encounter over the contemplation of the cosmos is that of the human spirit with the Spirit of the Creator! A Spirit truly divine, not a sort of "soul of the world" merged with the world as dreamed up by pantheism. The Universe of our experience itself rebels against this error: it speaks of a composed whole in spite of its dynamic unity and shows, apart from its beauties and undeniable harmonies, clear imperfections, irreconcilable with the divine fullness of Being. Divine Spirit, distinct and different from the world; not outside of the world, as if withdrawn in a disdainful solitude, abandoning its works to their destiny as deist theories assert; but on the contrary present in the world, as an almighty Creator, guardian, and lawgiver to whom the world is tied by an essential dependence in the heart of its being and its action. The divine Spirit reveals itself from the coldness of space to the scientist open to finding a purpose for the whole of existing reality: Spirit moved by a breath of goodness and love penetrating and explaining all which focuses and reveals itself, particularly in the human being made in its image and likeness. Because of this, this Spirit does not disdain from surrounding the human spirit with continuous and ineffable acts of love, such as the redemption accomplished

through its mysterious Incarnation. Similarly, the breadth of this conception of the Universe may have legitimately dethroned the ancient geocentric and anthropocentric ideas and thus, so to speak, shrunk our planet to the size of a grain of star dust and reduced man to the size of an atom on this speck of dust, relegating both to a corner of the Universe; but this does not represent (as some assert in speaking of the mystery of the Incarnation) an obstacle to

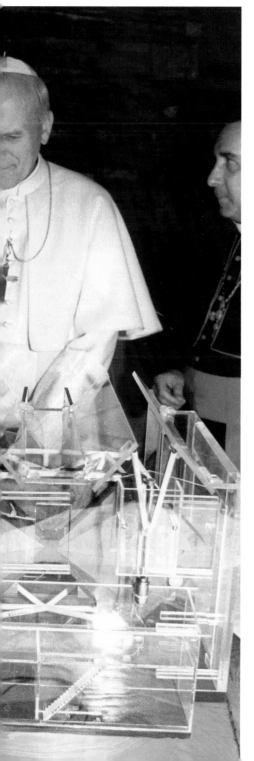

the love or the almightiness of He Who, being pure spirit, possesses an infinite superiority over matter, however great its cosmic dimensions in space, time, mass and energy.

As so, friends, above and beyond the deep respect which we entertain for all the sciences and for yours in particular, this is yet another reason why we are moved to pray: may the science of astronomy, founded on the highest and most universal horizons, the ideal of so many great men in the past such as Copernicus, Galileo, Kepler, and Newton, continue to bear the fruit of marvelous progress and, through to the heartfelt collaborations promoted by such groups as the International Astronomical Union, bring the astronomical vision of the Universe to an ever deeper perfection.

And finally, may the eternal light of God guide and enlighten you in your work, whose goal is to unveil the traces of His perfection and collect the echoes of His harmonies. Upon all those present we call heavenly favor, and as a token we invoke upon you our Apostolic Blessing.

As of 2009, we have seen much farther than the astronomers of 1952, and we have been able to refine many of the distance calculations the Pope quotes here. Our current best estimates for the dimensions of the Milky Way have not changed much since then — the distance from our Sun to the center of the Milky Way Galaxy is now determined to be 26,000 light-years — but new measurements place the Andromeda Galaxy a full 2.5 million light-years from us. The most distant objects seen by our telescopes are in the Hubble Ultra Deep Field image (shown on page 173); it contains an estimated 10,000 galaxies, the most distant being more than 13 thousand million light-years away, which means that their light was emitted just at most about 800 million years after

the Big Bang. And, of course, Pluto is no longer a considered a planet, but the first discovered representative of a different class of objects; the farthest "planet" of the Solar System is therefore Neptune, and on average light from Neptune takes about 4 hours and 10 minutes to reach us. But a whole family of trans-Neptunian objects (of which Pluto was the first discovered) have now been seen, out to nearly 100 times the Earth-Sun distance, or more than three times the distance to Neptune.

203

LETTER OF HIS HOLINESS JOHN PAUL II TO REVEREND GEORGE V. COYNE, S.J., DIRECTOR OF THE VATICAN OBSERVATORY (POPE JOHN PAUL II)

In 1987, the Pontifical Academy of Sciences held a Study Week in honor of the 300th anniversary of the publication of Newton's **Principia**. Following that event, Pope John Paul II prepared a letter for publication with the proceedings of that Study Week, addressed to the editor of those proceedings, Fr. George Coyne, S.J., the director of the Vatican Observatory. In keeping with that occasion, this letter outlined the Pope's thoughts about the relationship between science and religion in a more complete and systematic way than had ever been done before. It has become a classic description of why we do our science.

We reprint it here in full.

Opposite: Pope John Paul II examines a model of the Large Binocular Telescope with its designer, Dr. Roger Angel of the University of Arizona, and Fr. George Coyne, S.J., director of the Vatican Observatory, in Phoenix, Arizona, in 1987.

"Grace to you and peace from God our Father and the Lord Jesus Christ" (Ephesians 1:2).

As you prepare to publish the papers presented at the Study Week held at Castel Gandolfo on September 21-26, 1987, I take the occasion to express my gratitude to you and through you to all who contributed to that important initiative. I am confident that the publication of these papers will ensure that the fruits of that endeavor will be further enriched.

The three hundredth anniversary of the publication of Newton's *Philosophiae Naturalis Principia Mathematica* provided an appropriate occasion for the Holy See to sponsor a Study Week that investigated the multiple relationships among theology, philosophy, and the natural sciences. The man so honored, Sir Isaac Newton, had himself devoted much of his life to these same issues, and his reflections upon them can be found throughout his major works, his unfinished manuscripts, and his vast correspondence. The publication of your own papers from this Study Week, taking up again some of the same questions which this great genius explored, affords me the opportunity to thank you for the efforts you devoted to a subject of such paramount importance. The theme of your conference, "Our Knowledge of God and Nature: Physics, Philosophy, and Theology," is assuredly a crucial one for the contemporary world. Because of its impor-

tance, I should like to address some issues which the interactions among natural science, philosophy, and theology present to the Church and to human society in general.

The Church and the Academy engage one another as two very different but major institutions within human civilization and world culture. We bear before God enormous responsibilities for the human condition because historically we have had and continue to have a major influence on the development of ideas and values and on the course of human action. We both have histories stretching back over thousands of years: the learned, academic community dating back to the origins of culture, to the city and the

library and the school, and the Church with her historical roots in ancient Israel. We have come into contact often during these centuries, sometimes in mutual support, at other times in those needless conflicts which have marred

> ...Each discipline should continue to enrich, nourish, and challenge the other to be more fully what it can be and to contribute to our vision of who we are and who we are becoming.
> **Pope John Paul II**

Above: *As the feast of St. Ignatius Loyola, the founder of the Jesuit order, occurs on July 31 while the Pope is in residence in Castel Gandolfo, it is not unusual for the Popes to invite members of the Jesuit community to celebrate Mass with him on that day. Present here with His Holiness in 1994 were (from left) Sabino Maffeo, Gary Menard, José Funes, and Richard Boyle. At that time, Menard and Funes were scholastics, not yet ordained. Today, Fr. Menard is assistant headmaster at the Arrupe Jesuit High School in Denver, Colorado, and Fr. Funes is now director of the Vatican Observatory.*

both our histories. In your conference we met again, and it was altogether fitting that as we approach the close of this millennium we initiated a series of reflections together upon the world as we touch it and as it shapes and challenges our actions.

So much of our world seems to be in fragments, in disjointed pieces. So much of human life is passed in isolation or in hostility. The division between rich nations and poor nations continues to grow; the contrast between northern and southern regions of our planet becomes ever more marked and intolerable. The antagonism between races and religions splits countries into warring camps; historical animosities show no signs of abating. Even within the academic community, the separation between truth and values persists, and the isolation of their several cultures — scientific, humanistic, and religious — makes common discourse difficult if not at times impossible.

But at the same time we see in large sectors of the human community a growing critical openness towards people of different cultures and backgrounds, different competencies and viewpoints. More and more frequently, people are seeking intellectual coherence and collaboration, and are discovering values and experiences they have in common even within their diversities. This openness, this dynamic interchange, is a notable feature of the international scientific communities themselves, and is based on common interests, common goals and a common enterprise, along with a deep awareness that the insights and attainments of one are often important for the progress of the other. In a similar but more subtle way this has occurred and is continuing to occur among more diverse groups — among the communities that make up the Church, and even between the scientific community and the Church herself. This drive is essentially a movement towards the kind of unity which resists homogenization and relishes diversity. Such community is determined by a common meaning and by a shared understanding that evokes a sense of mutual involvement. Two groups which may seem initially to have nothing in common can begin to enter into community with one another by discovering a common goal, and this in turn can lead to broader areas of shared understanding and concern.

As never before in her history, the Church has entered into the movement for the union of all Christians, fostering common study, prayer, and discussions that "all may be one" (John 17:20). She has attempted to rid herself of every vestige of antisemitism and to emphasize her origins in and her religious debt to Judaism. In reflection and prayer, she has reached out to the great world religions, recognizing the values we all hold in common and our universal and utter dependence upon God.

Within the Church herself, there is a growing sense of "world church", so much in evidence at the last Ecumenical Council in which bishops native to every continent — no longer predominantly of European or even Western origin — assumed for the first time their common responsibility for the

....

Top: Present here with Pope John Paul II on the feast of St. Ignatius in 1995 are Br. Guy Consolmagno, Fr. William Stoeger, Fr. Sabino Maffeo, and Mr. David Brown. Brown, then a scholastic in the early stages of his Jesuit studies, only was visiting the Vatican Observatory that summer. He would finally join the observatory full-time upon completing his doctorate in astrophysics from Oxford in 2008.

entire Church. The documents from that Council and of the magisterium have reflected this new world-consciousness both in their content and in their attempt to address all people of good will. During this century, we have witnessed a dynamic tendency to reconciliation and unity that has taken many forms within the Church.

Nor should such a development be surprising. The Christian community in moving so emphatically in this direction is realizing in greater intensity the activity of Christ within her: "For God was in Christ, reconciling the world to himself" (2 Corinthians 5:19). We ourselves are called to be a continuation of the reconciliation of human beings, one with another and all with God. Our very nature as Church entails this commitment to unity.

Turning to the relationship between religion and science, there has been a definite, though still fragile and provisional, movement towards a new and more nuanced interchange. We have begun to talk to one another on deeper levels than before, and with greater openness towards one another's perspectives. We have begun to search together for a more thorough understanding of one another's disciplines, with their competencies and their limitations, and especially for areas of common ground. In doing so we have uncovered important questions which concern both of us, and which are vital to the larger human community we both serve. It is crucial

that this common search based on critical openness and interchange should not only continue but also grow and deepen in its quality and scope.

For the impact each has, and will continue to have, on the course of civilization and on the world itself, cannot be overestimated, and there is so much that each can offer the other. There is, of course, the vision of the unity of all things and all peoples in Christ, who is active and present with us in our daily lives — in our struggles, our sufferings, our joys and in our searchings — and who is the focus of the Church's life and witness. This vision carries with it into the larger community a deep reverence for all that is, a hope and assurance that the fragile goodness, beauty, and life we see in the universe is moving towards a completion and fulfilment which will not be overwhelmed by the forces of dissolution and death. This vision also provides a strong support for the values which are emerging both from our knowledge and appreciation of creation and of ourselves as the products, knowers, and stewards of creation.

The scientific disciplines too, as is obvious, are endowing us with an understanding and appreciation of our universe as a whole and of the incredibly rich variety of intricately related processes and structures which constitute its animate and inanimate components. This knowledge has given us a more thorough understanding of ourselves and of our humble yet unique role within creation. Through technology it also has given us the capacity to travel, to communicate, to build, to cure, and to probe in ways which would have been almost unimaginable to our ancestors. Such knowledge and power, as we have discovered, can be used greatly to enhance and improve our lives or they can be exploited to diminish and

destroy human life and the environment even on a global scale.

The unity we perceive in creation on the basis of our faith in Jesus Christ as Lord of the universe, and the correlative unity for which we strive in our human communities, seems to be reflected and even reinforced in what contemporary science is revealing to us. As we behold the incredible development of scientific research we detect an underlying movement towards the discovery of levels of law

> *Both religion and science must preserve their autonomy and their distinctiveness. Religion is not founded on science nor is science an extension of religion. Each should possess its own principles, its pattern of procedures, its diversities of interpretation and its own conclusions.*
> **Pope John Paul II**

and process which unify created reality and which at the same time have given rise to the vast diversity of structures and organisms which constitute the physical and biological, and even the psychological and sociological, worlds.

Contemporary physics furnishes a striking example. The quest for the unification of all four fundamental physical forces — gravitation, electromagnetism, the strong and weak nuclear interactions — has met with increasing success. This unification may well combine discoveries from the sub-atomic and the cosmological domains and shed light both on the origin of the universe and, eventually, on the origin of the laws and constants which govern its evolution. Physicists possess a detailed though incomplete and provisional knowledge of elementary particles and of the fundamental forces through which they interact at low and intermediate energies. They now have an acceptable theory unifying the electromagnetic and weak nuclear forces, along with much less adequate but still promising grand unified field theories which attempt to incorporate the strong nuclear interaction as well. Further in the line of this same development, there are already several detailed suggestions for the final stage, superunification, that is, the unification of all four fundamental forces, including gravity. Is it not important for us to note that in a world of such detailed specialization as contemporary physics there exists this drive towards convergence?

In the life sciences, too, something similar has happened. Molecular biologists have probed the structure of living material, its functions and its processes of replication. They have discovered that the same underlying constituents serve in the makeup of all living organisms on Earth and constitute both the genes and the proteins which these genes code. This is another impressive manifestation of the unity of nature.

By encouraging openness between the Church and the scientific communities, we are not envisioning a disciplinary unity between theology and science like that which exists within

Science can purify religion from error and superstition; religion can purify science from idolatry and false absolutes.
Pope John Paul II

a given scientific field or within theology proper. As dialogue and common searching continue, there will be growth towards mutual understanding and a gradual uncovering of common concerns which will provide the basis for further research and discussion. Exactly what form that will take must be left to the future. What is important, as we have already stressed, is that the dialogue should continue and grow in depth and scope. In the process we must overcome every regressive tendency to a unilateral reductionism, to fear, and to self-imposed isolation. What is critically important is that each discipline should continue to enrich, nourish and challenge the other to be more fully what it can be and to contribute to our vision of who we are and who we are becoming.

We might ask whether or not we are ready for this crucial endeavor. Is the community of world religions, including the Church, ready to enter into a more thorough-going dialogue with the scientific community, a dialogue in which the integrity of both religion and science is supported and the advance of each is fostered? Is the scientific community now prepared to open itself to Christianity, and indeed to all the great world religions, working with us all to build a culture that is more humane and in that way more divine? Do we dare to risk the honesty and the courage that this task demands? We must ask ourselves whether both

science and religion will contribute to the integration of human culture or to its fragmentation. It is a single choice and it confronts us all.

For a simple neutrality is no longer acceptable. If they are to grow and mature, peoples cannot continue to live in separate compartments, pursuing totally divergent interests from which they evaluate and judge their world. A divided community fosters a fragmented vision of the world; a community of interchange encourages its members to expand their partial perspectives and form a new unified vision.

Yet the unity that we seek, as we have already stressed, is not identity. The Church does not propose that science should become religion or religion science. On the contrary, unity always presupposes the diversity and the integrity of its elements. Each of these members should become not less itself but more itself in a dynamic interchange, for a unity in which one of the elements is reduced to the other is destructive, false in its promises of harmony, and ruinous of the integrity of its components. We are asked to become one. We are not asked to become each other.

To be more specific, both religion and science must preserve their autonomy and their distinctiveness. Religion is not founded on science nor is science an extension of religion. Each should possess its own principles, its pattern of procedures, its diversities of interpretation, and its own conclusions. Christianity possesses the source of its justification within itself and does not expect science to constitute its primary apologetic. Science must bear witness to its own worth. While each can and should support the other as distinct dimensions of a common human culture, neither ought to assume that it forms a necessary premise for the other. The unprecedented opportunity we have today is for a common interactive relationship in which each discipline

his Son Jesus Christ. And we are writing this that our joy may be complete" (1 John 1:3-4). Later the Church reached out to the sciences and to the arts, founding great universities and building monuments of surpassing beauty so that all things might be recapitulated in Christ (cf. Ephesians 1:10).

What, then, does the Church encourage in this relational unity between science and religion? First and foremost that they should come to understand one another. For too long a time they have been at arm's length. Theology has been defined as an effort of faith to achieve understanding, as *fides quaerens intellectum*. As such, it must be in vital interchange today with science just as it always has been with philosophy and other forms of learning. Theology will have to call on the findings of science to one degree or another as it pursues its primary concern for the human person, the reaches of freedom, the possibilities of Christian community, the nature of belief and the intelligibility of nature and history. The vitality and significance of theology for humanity will in a profound way be reflected in its ability to incorporate these findings.

Now this is a point of delicate importance, and it has to be carefully qualified. Theology is not to incorporate indifferently each new philosophical or scientific theory. As these findings become part of the intellectual culture of the time, however, theologians must understand them and test their value in bringing out from Christian belief some of the possibilities which have not yet been realized. The hylo-

Right: Faculty and students of the 2007 Vatican Observatory Summer School in audience with His Holiness Pope Benedict XVI.

The truth of the matter is that the Church and the scientific community will inevitably interact; their options do not include isolation.
Pope John Paul II

retains its integrity and yet is radically open to the discoveries and insights of the other.

But why is critical openness and mutual interchange a value for both of us? Unity involves the drive of the human mind towards understanding and the desire of the human spirit for love. When human beings seek to understand the multiplicities that surround them, when they seek to make sense of experience, they do so by bringing many factors into a common vision. Understanding is achieved when many data are unified by a common structure. The one illuminates the many: it makes sense of the whole. Simple multiplicity is chaos; an insight, a single model, can give that chaos structure and draw it into intelligibility. We move towards unity as we move towards meaning in our lives. Unity is also the consequence of love. If love is genuine, it moves not towards the assimilation of the other but towards union with the other. Human community begins in desire when that union has not been achieved, and it is completed in joy when those who have been apart are now united.

In the Church's earliest documents, the realization of community, in the radical sense of that word, was seen as the promise and goal of the Gospel: "That which we have seen and heard we proclaim also to you, so that you may have fellowship with us; and our fellowship is with the Father and with

morphism of Aristotelian philosophy, for example, was adopted by the medieval theologians to help them explore the nature of the sacraments and the hypostatic union. This did not mean that the Church adjudicated the truth or falsity of the Aristotelian insight, since that is not her concern. It did mean that this was one of the rich insights offered by Greek culture, that it needed to be understood and taken seriously and tested for its value in illuminating various areas of theology. Theologians might well ask, with respect to contemporary science, philosophy, and the other areas of human knowing, if they have accomplished this extraordinarily difficult process as well as did these medieval masters.

If the cosmologies of the ancient Near Eastern world could be purified and assimilated into the first chapters of Genesis, might not contemporary cosmology have something to offer to our reflections upon creation? Does an evolutionary perspective bring any light to bear upon theological anthropology, the meaning of the human person as the *imago Dei*, the problem of Christology — and even upon the development of doctrine itself? What, if any, are the eschatological implications of contemporary cosmology, especially in light of the vast future of our universe? Can theological method fruitfully appropriate insights from

*Galileo
saw nature
as a book
whose author is God
in the same way
that Scripture
has God
as its author.*
Pope Benedict XVI

scientific methodology and the philosophy of science?

Questions of this kind can be suggested in abundance. Pursuing them further would require the sort of intense dialogue with contemporary science that has, on the whole, been lacking among those engaged in theological research and teaching. It would entail that some theologians, at least, should be sufficiently well-versed in the sciences to make authentic and creative use of the resources that the best-established theories may offer them. Such an expertise would prevent them from making uncritical and overhasty use for apologetic purposes of such recent theories as that of the "Big Bang" in cosmology. Yet it would equally keep them from discounting altogether the potential relevance of such theories to the deepening of understanding in traditional areas of theological inquiry.

In this process of mutual learning, those members of the Church who are themselves either active scientists or, in some special cases, both scientists and theologians could serve as a key resource. They can also provide a much-needed ministry to others struggling to integrate the worlds of science and religion in their own intellectual and spiritual lives, as well as to those who face difficult moral decisions in matters of technological research and application. Such bridging ministries must be nurtured and encouraged. The Church long ago recognized the importance of

such links by establishing the Pontifical Academy of Sciences, in which some of the world's leading scientists meet together regularly to discuss their researches and to convey to the larger community where the directions of discovery are tending. But much more is needed.

The matter is urgent. Contemporary developments in science challenge theology far more deeply than did the introduction of Aristotle into Western Europe in the thirteenth century. Yet these developments also offer to theology a potentially important resource. Just as Aristotelian philosophy, through the ministry of such great scholars as St. Thomas Aquinas, ultimately came to shape some of the most profound expressions of theological doctrine, so can we not hope that the sciences of today, along with all forms of human knowing, may invigorate and inform those parts of the theological enterprise that bear on the relation of nature, humanity, and God?

Can science also benefit from this interchange? It would seem that it should. For science develops best when its concepts and conclusions are integrated into the broader human culture and its concerns for ultimate meaning and value. Scientists cannot, therefore, hold themselves entirely aloof from the sorts of issues dealt with by philosophers and theologians. By devoting to

....

Right: Fernando Comerón, an expert on star formation and brown dwarfs at the European Southern Observatory, was an alumnus of the 1990 Vatican Observatory Summer School and one of the instructors at the 2007 school. Here he introduces his daughter Ines to His Holiness Pope Benedict XVI at the Pope's audience with the 2007 school.

these issues something of the energy and care they give to their research in science, they can help others realize more fully the human potentialities of their discoveries. They can also come to appreciate for themselves that these discoveries cannot be a genuine substitute for knowledge of the truly ultimate. Science can purify religion from error and superstition; religion can purify science from idolatry and false absolutes. Each can draw the other into a wider world, a world in which both can flourish.

For the truth of the matter is that the Church and the scientific community will inevitably interact; their options do not include isolation. Christians will inevitably assimilate the prevailing ideas about the world, and today these are deeply shaped by science. The only question is whether they will do this critically or unreflectively, with depth and nuance or with a shallowness that debases the Gospel and leaves us ashamed before history. Scientists, like all human beings, will make decisions upon what ultimately gives meaning and value to their lives and to their work. This they will do well or poorly, with the reflective depth that theological wisdom can help them attain, or with an unconsidered absolutizing of their results beyond their reasonable and proper limits.

Both the Church and the scientific community are faced with such inescapable alternatives. We shall make our choices much better if we live in a collaborative interaction in which we are called continually to be more. Only a dynamic relationship between theology and science can reveal those limits which support the integrity of either discipline, so that theology does not profess a pseudo-science and science does not become an unconscious theology. Our knowledge of each other can lead us to be more authentically ourselves. No one can read the history of the past century and not realize that crisis is upon us both. The uses of science have on more than one occasion

ADDRESS TO THE 2007 VATICAN OBSERVATORY SUMMER SCHOOL (POPE BENEDICT XVI)

Pope Benedict XVI addressed these comments to the students of the 11th Vatican Observatory Summer School in Astrophysics in June 2007. While intended primarily for those 26 students from around the world, they also express his devotion to our work and the hopes that come with the participation of the Vatican in the wider world of astronomy.

proved massively destructive, and the reflections of religion have too often been sterile. We need each other to be what we must be, what we are called to be.

And so on this occasion of the Newton Tercentennial, the Church speaking through my ministry calls upon herself and the scientific community to intensify their constructive relations of interchange through unity. You are called to learn from one another, to renew the context in which science is done and to nourish the inculturation which vital theology demands. Each of you has everything to gain from such an interaction, and the human community which we both serve has a right to demand it from us.

Upon all who participated in the Study Week sponsored by the Holy See and upon all who will read and study the papers herein published I invoke wisdom and peace in our Lord Jesus Christ and cordially impart my Apostolic Blessing.

From the Vatican, June 1, 1988.
Joannes Paulus P.P. II

I am pleased to greet the faculty and students of the Eleventh Vatican Observatory Summer School, and I thank the Director, Father José Funes, for his kind words of greeting in your name.

Since its establishment in 1891, the Vatican Observatory has sought to demonstrate the Church's desire to embrace, encourage and promote scientific study, on the basis of her conviction that "faith and reason are like two wings on which the human spirit rises to the contemplation of truth" (*Fides et Ratio, Proemium*). The Jesuit Fathers and Brothers who staff the Observatory are not only involved in astronomical research, but are also committed to offering educational opportunities for the next generation of astronomers. The Vatican Observatory Summer School is

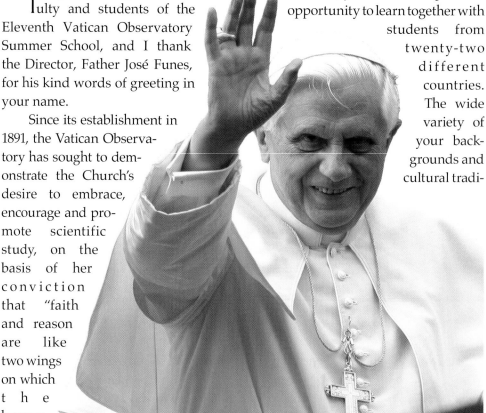

a concrete sign of that commitment.

Your program this month is devoted to the study of Extrasolar Planets. In addition to your demanding research, however, you will have a precious opportunity to learn together with students from twenty-two different countries. The wide variety of your backgrounds and cultural tradi-

tions can be a source of great enrichment to you all. I encourage you to make the most of this experience, and I offer my prayerful good wishes that your small international community may become a promising sign of greater scientific collaboration for the benefit of the entire human family.

In the days to come, may you find spiritual consolation in the study of the stars that "shine to delight their Creator" (Bar 3:35). Upon you and your families I cordially invoke God's blessings of wisdom, joy and peace.

Pope Benedict XVI,
June 11, 2007

ADDRESS TO THE PONTIFICAL ACADEMY OF SCIENCES, 2008 (POPE BENEDICT XVI)

*P*ope *Benedict touched on a number of issues involving both biology and astronomy in his address to the plenary session of the PAS in October 2008. In the following address, three points are notable. First, in much the same way expressed in Fr. Stoeger's article about Creation and the Big Bang, the Pope draws a clear distinction between God as primary cause in the realm of the life sciences, and the function of those sciences to describe "not creation, but rather a mutation or transformation" of that which has been created. (One also hears an echo of his comments on the role of mathematics in Fr. Whitman's chapter.) Of course, the basic theology can be found in Aquinas — hearkening back to the very first essay in this chapter, from Pope Leo XIII. Second, it is interesting to see how Pope Benedict XVI quotes Galileo as an authority on the nature of the relationship between faith and science. And finally, note how the trend discussed at the beginning of this chapter has become complete. Where, a hundred years ago, the Church viewed science as an apologetic tool supporting well-established doctrine, now with Pope Benedict XVI we also see the scientific understanding of Creation as itself providing the knowledge to enrich our theological understanding of the Creator.*

The laws of nature ... are a great incentive to contemplate the works of the Lord with gratitude.
Pope Benedict XVI

In choosing the topic Scientific Insight into the Evolution of the Universe and of Life, you seek to focus on an area of enquiry which elicits much interest. In fact, many of our contemporaries today wish to reflect upon the ultimate origin of beings, their cause and their end, and the meaning of human history and the universe.

In this context, questions concerning the relationship between science's reading of the world and the reading offered by Christian Revelation naturally arise. My predecessors Pope Pius XII and Pope John Paul II noted that there is no opposition between faith's understanding of creation and the evidence of the empirical sciences. Philosophy in its early stages had proposed images to explain the origin of the cosmos on the basis of one or more elements of the material world. This genesis was not seen as a creation, but rather a mutation or transformation; it involved a somewhat horizontal interpretation of the origin of the world. A decisive advance in understanding the origin of the cosmos was the consideration of being *qua* being and the concern of metaphysics with the most basic question of the first or transcendent origin of participated being. In order to develop and evolve, the world must first be, and thus have come from nothing into being. It must be created, in other words, by the first Being who is such by essence.

To state that the foundation of the cosmos and its developments is the provident wisdom of the Creator is not to say that creation has only to do with the beginning of the history of the world and of life. It implies, rather, that the Creator founds these developments and supports them, underpins them and sustains them continuously. Thomas Aquinas taught that the notion of creation must transcend the horizontal origin of the unfolding of events, which is history, and consequently all our purely naturalistic ways of thinking and speaking about the evolution of the world. Thomas observed that creation is neither a movement nor a mutation. It is instead the foundational and continuing relationship that links the creature to the Creator, for he is the cause of every being and all becoming (cf. *Summa Theologiae*, I, q. 45, a. 3).

To "evolve" literally means "to unroll a scroll," that is, to read a book. The imagery of nature as a book has its roots in Christianity and has been held dear by many scientists. Galileo saw nature as a book whose author is God in the same way that Scripture has God as its author. It is a book whose history, whose evolution, whose "writing" and meaning, we "read" according to the different approaches of the sciences, while all the time presupposing the foundational presence of the author who has wished to reveal himself therein. This image also helps us to understand that the world, far from originating out of chaos, resembles an ordered book; it is a cosmos. Notwithstanding elements of the irrational, chaotic, and the destructive in the long processes of change in the cosmos, matter as such is "legible." It has an inbuilt "mathematics." The human mind therefore can engage not only in a "cosmography" studying measurable phenomena but also in a "cosmology" discerning the visible inner logic of the cosmos. We may not at first be able to see the harmony both of the whole and of the relations of the individual parts, or their relationship to

divine Truth, can help philosophy and theology to understand ever more fully the human person and God's Revelation about man, a Revelation that is completed and perfected in Jesus Christ. For this important mutual enrichment in the search for the truth and the benefit of mankind, I am, with the whole Church, profoundly grateful".

> There is a special concept of the cosmos in Christianity.
> **Pope Benedict XVI**

the whole. Yet, there always remains a broad range of intelligible events, and the process is rational in that it reveals an order of evident correspondences and undeniable finalities: in the inorganic world, between microstructure and macrostructure; in the organic and animal world, between structure and function; and in the spiritual world, between knowledge of the truth and the aspiration to freedom. Experimental and philosophical inquiry gradually discovers these orders; it perceives them working to maintain themselves in being, defending themselves against imbalances, and overcoming obstacles. And thanks to the natural sciences we have greatly increased our understanding of the uniqueness of humanity's place in the cosmos.

The distinction between a simple living being and a spiritual being that is *capax Dei*, points to the existence of the intellective soul of a free transcendent subject. Thus the Magisterium of the Church has constantly affirmed that "every spiritual soul is created immediately by God — it is not 'produced' by the parents — and also that it is immortal" (*Catechism of the Catholic Church*, 366). This points to the distinctiveness of anthropology, and invites exploration of it by modern thought.

Distinguished Academicians, I wish to conclude by recalling the words addressed to you by my predecessor Pope John Paul II in November 2003: "Scientific truth, which is itself a participation in

Upon you and your families, and all those associated with the work of the Pontifical Academy of Sciences, I cordially invoke God's blessings of wisdom and peace.

October 31, 2008

Pope Benedict XVI became one of the first world leaders to salute the International Year of Astronomy. Speaking at his weekly prayer of the Angelus at St. Peter's Square on the winter solstice, 2008, he summarized the long history of astronomy and the Popes:

From **ADDRESS AT
THE WEEKLY ANGELUS
(POPE BENEDICT XVI)**

Beyond its historical dimension, this mystery of salvation also has a cosmic dimension: Christ is the sun of grace who, with his life, "transfigures and enflames the expectant universe" (cf. Liturgy). The Christmas festivity is placed within and linked to the winter solstice when, in the north-

ern hemisphere, the days begin once again to lengthen. In this regard perhaps not everyone knows that in St. Peter's Square there is also a meridian; in fact, the great obelisk casts its shadow in a line that runs along the paving stones toward the fountain beneath this window and in these days, the shadow is at its longest of the year. This reminds us of the role of astronomy in setting the times of prayer. The Angelus, for example, is recited in the morning, at noon, and in the evening, and clocks were regulated by the meridian which in ancient times made it possible to know the "exact midday."

The fact that the winter solstice occurs exactly today, December 21, and at this very time, offers me the opportunity to greet all those who will be taking part in various capacities in the initiatives for the World Year of Astronomy, 2009, established on the fourth centenary of Galileo Galilei's first observations by telescope. Among my Predecessors of venerable memory there were some who studied this science, such as Sylvester II who taught it, Gregory XIII to whom we owe our calendar, and St. Pius X who knew how to build sundials. If the heavens, according to the Psalmist's beautiful words, "are telling the glory of God" (Psalm 19[18]:1), the laws of nature which over the course of centuries many men and women of science have enabled us to understand better are a great incentive to contemplate the works of the Lord with gratitude.

*St. Peter's Square
Fourth Sunday of Advent,
December 21, 2008*

From **HOMILY ON
THE SOLEMNITY
OF THE EPIPHANY
OF THE LORD
(POPE BENEDICT XVI)**

In his address to the Pontifical Academy of Sciences, above, Pope Benedict quoted his predecessor John Paul II, that "scientific truth ... can help philosophy and theology to understand ever more fully the human person and God's Revelation about man." Two months after that address, in his homily on the feast of the Wise Men, he showed how this can be done. Here, theology and astronomy are woven together into a beautiful image of the Incarnation:

In this year 2009, which has been dedicated in a special way to astronomy to mark the fourth centenary of Galileo Galilei's first observations with the telescope, we cannot fail to pay particular attention to the symbol of the star that is so important in the Gospel account of the Magi (cf. Matthew 2:1-12). In all likelihood the Wise Men were astronomers. From their observation point, situated in the East compared to Palestine, perhaps in Mesopotamia, they had noticed the appearance of a new star and had interpreted this celestial phenomenon as the announcement of the birth of a king, specifically that in accordance with the Sacred Scriptures of the King of the Jews (cf. Numbers 24:17). The Fathers of the Church also saw this unique episode recounted by St. Matthew as a sort of cosmic "revolution" caused by the Son of God's entry into the world. For example, St. John Chrysostom writes: "The star, when it stood over the young Child, stayed its course again: which thing itself was of a greater power than belongs to a star, now to hide itself, now to appear, and having appeared to stand still" (Homily on the Gospel of Matthew 7,

3). St. Gregory of Nazianzen states that the birth of Christ gave the stars new orbits (cf. *Dogmatic Poems*, v, 53-64: PG 37, 428-429). This is clearly to be understood in a symbolic and theological sense. In effect, while pagan theology divinized the elements and forces of the cosmos, the Christian faith, in bringing the biblical Revelation to fulfillment, contemplates only one God, Creator and Lord of the whole universe.

The divine and universal law of creation is divine love, incarnate in Christ. However, this should not be understood in a poetic but in a real sense. Moreover, this is what Dante himself meant when, in the sublime verse that concludes the *Paradiso* and the entire *Divina Commedia*, he describes God as "the Love which moves the sun and the other stars" (*Paradiso*, XXXIII, 145). This means that the stars, planets and the whole universe are not governed by a blind force, they do not obey the dynamics of matter alone. Therefore, it is not the cosmic elements that should be divinized. Indeed, on the contrary, within everything and at the same time above everything there is a personal will, the Spirit of God, who in Christ has revealed himself as Love (cf. Encyclical *Spe Salvi*, 5). If this is the case, then as St. Paul wrote to the Colossians people are not slaves of the "elemental spirits of the universe" (cf. Colossians 2:8) but are free, that is, capable of relating to the creative freedom of God. God is at the origin of all things and governs all things, not as a cold and

anonymous engine but rather as Father, Husband, Friend, Brother, and as the Logos, "Word-Reason" who was united with our mortal flesh once and for all and fully shared our condition, showing the superabundant power of his grace. Thus there is a special concept of the cosmos in Christianity which found its loftiest expression in medieval philosophy and theology. In our day too, it shows interesting signs of a new flourishing, thanks to the enthusiasm and faith of many scientists who following in Galileo's footsteps renounce neither reason nor faith; instead they develop both in their reciprocal fruitfulness.

Christian thought compares the cosmos to a "book"; the same Galileo said this as well, considering it as the work of an Author who expresses himself in the "symphony" of the Creation. In this symphony is found, at a certain point, what might be called in musical terminology a "solo," a theme given to a single instrument or voice; and it is so important that the significance of the entire work depends on it. This "solo" is Jesus, who is accompanied by a royal sign: the appearance of a new star in the firmament. Jesus is compared by ancient Christian writers to a new sun. According to current astrophysical knowledge, we should compare it with a star that is even more central, not only for the Solar System but also for the entire known universe. Within this mysterious design simultaneously physical and metaphysical, which led to the appearance of the human being as the crowning of Creation's elements Jesus came into the world:

"born of woman" (Galatians 4:4), as St. Paul writes. The Son of man himself epitomizes the earth and Heaven, the Creation and the Creator, the flesh and the Spirit. He is the center of the cosmos and of history, for in him the Author and his work are united without being confused with each other.

St. Peter's Basilica
Tuesday, January 6, 2009

> *The stars, planets and the whole universe ... do not obey the dynamics of matter alone.... God is at the origin of all things and governs all things, not as a cold and anonymous engine but rather as Father, Husband, Friend, Brother, and as the Logos.*
> **Pope Benedict XVI**

Texts of the Papal speeches to the PAS reprinted here are taken from Papal Addresses to the Pontifical Academy of Sciences, *compiled and edited by Marcelo Sánchez Sorondo (Vol. 100, Pontificiae Academiae Scientiarum Varia, Vatican City, 2003; 524 pp.).*

QUESTIONS
ABOUT STARS

Above: The Vatican Advanced Technology Telescope (left) and the Sub-Millimeter Telescope as viewed in winter from the dome of the Large Binocular Telescope at the Mt. Graham International Observatory. Photo by Alex Lovell-Troy.

ABOUT THE UNIVERSE

• Fr. Christopher CORBALLY, s.j., and
Br. Guy CONSOLMAGNO, s.j. •

Above: Fr. Georges Lemaître, Belgian priest and astrophysicist, was one of the foundations of what came to be known as the "Big Bang" theory. Credit: "Archives Lemaître" Université catholique de Louvain. Institut d'Astronomie et de Géophysique G. Lemaître. Louvain-la-Neuve. Belgique.

How can we know when the universe began? Is it true that it has been calculated to be 13 billion years ago? When did the first stars appear?

The evidence that there was some sort of primordial state of the universe which expanded (via what is popularly called the "Big Bang") into what we observe today is far more extensive than we can possibly do justice to here. But two fundamental points are these: we observe clusters of galaxies everywhere in the universe moving away from each other exactly as such a theory predicts; and we can observe the universe filled, in every direction, with microwave radiation that also exactly fits what the theory predicts. (The Big Bang theory makes other predictions that have also been confirmed, time and again.)

By observing the motions of these galaxies and working "backwards" to calculate the time when they must have all been together at one point, one concludes that this primal expansion of the universe has been going on for 13.7 billion years.

What happened, then, 13.7 billion years ago? That's where things get very tricky, and our theories — both physical and philosophical — become much more speculative. This is discussed by Fr. Stoeger in his chapter, on page 174.

Determining when the first stars appeared is still a hot topic for people studying the physics of the Big Bang and star formation. Recent theories suggest that stars could have formed as early as 200 million years after the Big Bang. But this age is certain to be refined by future work.

How do we know the overall shape of the universe? Is it curved?

Einstein's Theory of General Relativity, proposed in 1916, outlined how the force of gravity can be described mathematically as a "curvature" of space. We could imagine how the two-dimensional surface of a piece of paper could be "curved" by being bent in a third dimension, but it is much harder to visualize how our three dimensions of space plus the dimension of time are "curved." One way to see what is meant is to notice that in "flat" space an object in motion would continue to travel in a constant direction, a straight line, absent any other force acting on it, but an object in orbit around a star is moving in a curved way. Einstein suggested that this curved path could be interpreted as the effect of the star's gravity curving space in the way that is traced out by the orbiting planet.

Space is filled with stars, gathered into galaxies and clusters of galaxies. All the matter in these stars presumably curves, or warps, space in such a way that the motions of every galaxy should eventually crash them into each other. If space were eternal and infinite, as people thought back in 1916, then why hadn't this already happened? To put it another way, what prevents all the mass of the universe from falling into itself?

Einstein proposed that another factor, which he called the *cosmological constant*, must exist to counter this curvature. However, in 1922, the Russian physicist Alexander Friedmann suggested that the universe can be expanding, in a way he related to the pos-

Top: Stephan's Quintet is a compact group of five galaxies in the constellation of Pegasus discovered by Édouard Stephan in 1877. This image was taken at the VATT by Matt Nelson, University of Virginia.

sible curvatures of space: if a universe started out with a sufficiently large expanding velocity, it can continue to expand indefinitely, even against the force of gravity. A 1927 paper by the Belgian astrophysicist and Catholic priest Georges Lemaître proposed an entire cosmology based on such an expansion of the universe from a single highly dense quantum state.

By examining the motions of distant clusters of galaxies and looking for variations within the cosmic radiation left over from the initial highly dense, energetic state of the universe (the Big Bang) one can actually measure the overall curvature of space that these galaxies and radiation are traversing. If the curvature is *positive* (picture the piece of paper curved back into itself like the surface of a sphere), then eventually the expansion of the universe should stop and two galaxy clusters that originally were moving apart would begin to fall towards one another and eventually meet each other again on the other side of the "sphere." If the curvature is *negative* (picture the piece of paper curved like a saddle), then the

clusters of galaxies never meet. But all the measurements to date indicate that the overall curvature is neither positive nor negative, but *flat*.

Even if the universe has positive curvature, the cosmological constant could cause the galaxies to expand away from one another, if it is big enough to overcome the other components causing the positive curvature. Once this occurs in an expanding universe, it will continue forever — unless the physics involved is even stranger than we've assumed!

Is it true that the universe is expanding faster now than it did when it was first formed, soon after the "Big Bang"? What is the significance of this phenomenon?

In 1929, the American astronomer Edwin Hubble observed that clusters of galaxies showed exactly the kind of motion predicted by the theory of an expanding universe. Because light travels at a finite speed, what we see today from distant objects is in fact light emitted from them a very long time ago, and thus one can actually look "back in time" to earlier epochs in the universe to constrain these theories. About 10 years ago, astronomers making careful measurements of the expansion speed of the most distant galaxies discovered that the expansion rate today had not slowed down compared to that in the past, as one might have expected if the mutual attraction of all the universe's gravity were countering the initial expansion velocity, but actually the opposite is true: the expansion of the universe is accelerating. Apparently Lemaître and Einstein were both correct: the universe not only started with a very energetic big bang, but it also has some sort of energy (which can be expressed as Einstein's "cosmological constant") to accelerate the expansion of the universe.

This "dark energy" has many significant implications for our understanding of the universe. For one thing, it allows us to understand how the universe can be full of mass, yet still have a "flat" curvature. But perhaps more exciting is simply the recognition that it exists. Until we made measurements of the curvature of the universe and its expansion rate, there was little reason to suspect the existence of dark energy; but now our best theories to date suggest that it actually represents three-quarters of all the "stuff" (mass/energy) in the universe!

Is the universe infinite, or does it have a boundary?

Our best understanding of the moment suggests that, in an odd way, both statements may be true. We certainly know that, given the observed expansion of the universe where the farther away we look, the faster the galaxies appear to recede from us (which is what would be expected for a universe that was expanding uniformly), we cannot observe anything beyond a horizon where the expansion appears to be moving from us at the speed of light. For a universe 13.7 billion years old, this horizon sits 13.7 billion light-years from us. So that is, in one sense, one boundary of the observable universe.

More detailed models of how the Big Bang proceeded suggest that very early in its history it may have suddenly "inflated" such that material that was originally within our horizon, and thus able to affect the material we can still see today, was pushed beyond that horizon. In addition, observations in the past 10 years have shown that the expansion of the universe is actually accelerating, which means that galaxy clusters at the edge of our "horizon" will eventually pass beyond that horizon, and that we must have lost touch with distant galaxy clusters over the age of the universe. Adding all these factors together suggests that material once in contact with our part of the universe, and thus in principle "knowable" by observing how its presence once among us affected what we still can see today, now extends more than 150 billion light-years away from us. This would mark a bigger boundary to the (at least in some way knowable) universe.

But none of this rules out the possibility that there could be more to the universe even beyond that boundary. It merely says that, so far as it would have any effect on what we can measure or

....

Top: Magnetometer, an instrument to measure the Earth's magnetic field, made by Carpentier (Paris) and donated to the Vatican Observatory by Fr. Lais in 1891.

calculate now, we can make no statement at all about the existence of such material or not. The universe could be finite or infinite, for all we could ever know.

Can one think of "space" outside the universe?

Only by limiting what you mean by "universe." The important thing to remember, though it is hard to understand, is that "space" and "time" are intimately connected (according to Einstein's Theory of General Relativity, our best theory to date), and when we speak of the universe expanding, we do not mean material moving into an otherwise empty void but rather the space and time of the void itself expanding. Space and time itself begins at the moment when the universe begins (if one can speak of such a beginning as a "moment"). There is no "outside" to this space and time.

Some cosmologists have postulated that other universes could exist, but they would not be some "place" or "time" different from our own universe. They would have to exist in a different dimension, or a different way, than our universe exists.

Another suggestion is the existence of an extremely large number of cosmic domains, each possessing its own space-time geometry within an overarching mega-universe. The many individual domains — you could think of them as "universes" since they do not interact with each other, existing beyond the 150 billion light-year horizon described above — could develop as "bubbles" within this larger mega-universe, expanding or contracting and possessing their own space-time structures.

How many galaxies are there in the universe? Approximately how

many stars are there in the largest galaxy that we know of? And how many stars are in the smallest galaxy? Is it possible to guess from this how many stars there are in the universe?

The famous Italian physicist Enrico Fermi, who worked in his later years at the University of Chicago, used to like to challenge physics students in oral exams with the question, "How many piano tuners are there in Chicago?" When the student would look totally baffled at such a question in a physics exam, Fermi would lead them through the way an astronomer would approach such a question. (Roughly how many people live in Chicago? About how many households do they live in? What fraction of these houses have pianos? How often is a typical piano tuned? How many pianos can one tuner tune in a day? Thus, how many tuners would you need to service all those pianos in Chicago?) The answer would be only approximate, maybe 10 times too big or too small, but for astronomers such figures are at least places to start to work out better theories.

And so, asking how many stars there are in the visible universe is like estimating the number of piano tuners. We don't expect to come up with a really accurate number. But we should be able to get an idea of how big such a number would be.

First, recognize that we can only speak of the "visible" universe, within our own "horizon" of material that is moving away from us at a speed slower than the speed of light and so able to send light to our telescopes. The simplest way to estimate the number of galaxies is to use a telescope like Hubble to image individual galaxies in a tiny portion of the sky (see the image on page 173), and then — assuming the number is roughly the same in every direction — calculate from that how many galaxies it would take to fill the

whole sky. In 1999, one such estimate came up with a count of 125 billion galaxies.

However, better technology is now showing us farther and fainter galaxies, at least doubling that number, or about 250 billion galaxies. And some more distant galaxies may not be visible to Hubble but require infrared or radio telescopes bigger than we have available at the moment. So this should be considered just a lower limit: rather, we should say that we can see *at least* 250 billion galaxies.

Among galaxies close enough to us, we can estimate the number of their stars simply by measuring their total brightness and then assuming an average brightness per star. The largest galaxies are Giant Elliptical Galaxies. One such galaxy, known as Markarian 348, has been estimated to have up to 100 trillion stars. By contrast, a small galaxy near our own Milky Way, Wilman 1, has been estimated to have only about 500,000 stars, smaller than some of the globular clusters within our own Galaxy.

But typically, we expect an average galaxy to have about 100 billion stars. Thus, if 100 billion (one followed by 11 zeros) galaxies have 100 billion stars, that comes to 10 thousand billion billion (one followed by 22 zeros), or 10 sextillion stars. That, of course, is a low estimate since we know there may be two to five to ten times more galaxies than merely 100 billion. But that does give an idea of the size of the visible universe!

What are the fundamental elements of matter and energy that the universe is made from?

Our best understanding today is that there are three fundamental types of material in the universe. The best known is ordinary matter, called "baryonic" matter: atoms

and things made of atoms, like stars and planets. But this appears to make up only 4 percent of the universe.

By observing the orbits of stars in nearby galaxies, we have learned that there is much more mass in these galaxies than can be accounted for just by the visible stars; this material has become known as "dark matter," and it is thought to make up another 21 percent of the universe.

But the way in which the whole universe appears to be expanding has led us in recent years to postulate the existence of a third element to the universe. We call it "dark energy" — energy, because it is apparently causing the universe's expansion to accelerate; and dark, because we're completely in the dark about the nature of this energy!

The most likely candidate for this dark energy is the energy represented by the cosmological constant described above. This is the same as "vacuum energy," energy that is not due to matter particles, but rather to the lowest energy state of fields: even in a vacuum there is always a certain amount of energy. This energy density remains constant, despite the continual expansion of the universe, and induces a repulsive gravitational force leading to the acceleration of the expansion of the universe.

The elements of the universe are discussed further in Fr. Omizzolo's chapter, on page 102. The philosophical implications of the difference between "vacuum" and "nothing" are discussed by Fr. Stoeger on page 174.

In addition to these major components, the background radiation observed by radio telescopes is another element of the universe. Today it is negligible, but in the early moments after the Big Bang, it was the dominant component of the universe, from which the other components eventually were formed. As the universe expands, its density (and hence importance) decreases. ●

CLOSER TO EARTH

Is it theoretically possible for humans to travel out of the Solar System, or even out of our Galaxy, like you see in science fiction movies?

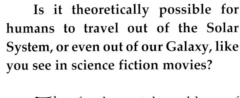

The fundamental problem of space travel is the distance between the places we want to visit. Travel to other planets within our own Solar System is certainly possible. We were able to visit the Moon 40 years ago, with 40-year-old technology. But while we could get to the Moon and return in eight days, a round-trip voyage to Mars would take about a hundred times as long, more than two years, to complete. We know in theory how to make such a trip, but actually building a system big enough and reliable enough to send human beings would challenge our technology today. About half the robot missions we have sent to Mars so far have failed due to technical problems; and none of them have been able to return samples back to Earth.

If going to Mars is a hundred times harder than going to the Moon, leaving the Solar System entirely is a hundred million times worse: the distance from Earth to Proxima Centauri, the star nearest to us (after the Sun) is more than a hundred million times the distance from Earth to the Moon. If you drew a map where the distance from Earth to the Moon was only one centimeter, you'd need a piece of paper a thousand kilometers wide to

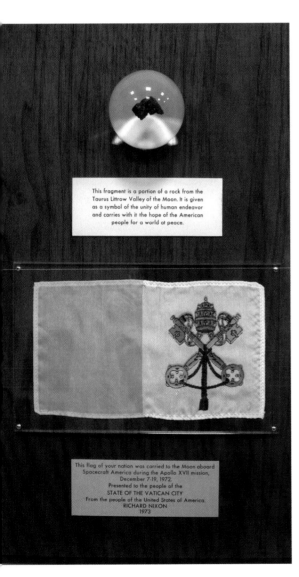

Above: A fragment of the Moon retrieved from the Taurus-Littrow valley by Apollo 17 astronauts in 1972, and a Vatican flag taken by them to the Moon, given to the Vatican in 1973.

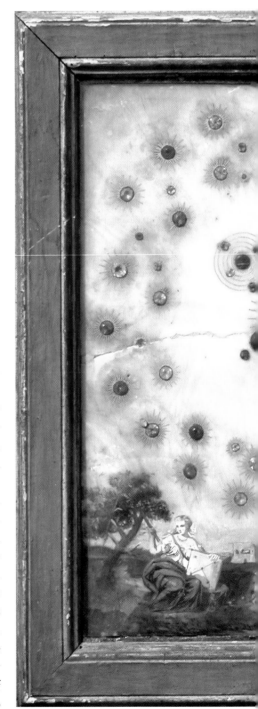

map the distance to Proxima Centauri. Traveling at the speed of the Apollo spacecraft to the Moon, a trip to that star would take a million years.

And that is just the nearest star. The nearest galaxy is a million times farther away — our star map would now have to stretch from here to Jupiter!

That is why science fiction stories rely on gimmicks like "warp drives" to power their space ships and skip over these immense distances. Those authors who even bother to explain how such drives work usually just wave their hands about "bending space" without seriously considering all the scientific implications. We have no idea if such a drive is even possible, much less within the realm of any conceivable technology.

It may not be impossible — not long ago, trips to the Moon were considered science fiction, too — but trips to other stars won't be happening any time soon.

Do we know of any other solar systems in the universe? Are they similar to ours?

Yes, as of this writing we know of hundreds of stars that have planets; around many of them, more than one planet has already been discovered. Because of the limits of the way we search for such planets, we have not been able to see any Earth-like planets yet, but that is only a matter of time and improved detection techniques. Fr. Koch writes about the search for other solar systems on page 162.

Can the activity of our Sun be the reason for the existence of global warming?

One of the very practical reasons for studying other planets is that in this way we are better able to understand our Earth and its place in the solar system. The scientific issues of global warming are so complex that having other planets to look at, to test our theories for what is happening on Earth, is a necessary part of any understanding of the problem.

Here is what we know: Venus has a very thick carbon dioxide atmosphere, and its surface temperature is far hotter than can be explained simply by it being closer to the Sun than we are. Mars was also clearly once much warmer than it is today (we see evidence of dried up river beds, whereas now it is much too cold for water to be liquid on its surface), and our best evidence is that this occurred when it also had a thicker atmosphere with "greenhouse" gases to hold in the heat. Indeed, the temperature of Earth today is already elevated to some degree by natural greenhouse gases like carbon dioxide and methane. We also know that the amount of carbon dioxide in our atmosphere today is about 35 percent higher than it was at the beginning of the industrial revolution, 200 years ago. The exact details of how this carbon dioxide will affect our temperature is still not completely solved, but certainly it must have some effect.

Yet we also know that other factors, such as small changes in the Sun's brightness or changes in the shape and orientation of Earth's orbit, can also affect climate. In the past, Earth has been both hotter and cooler than today because of these effects.

Certainly all of these factors affect global warming. We cannot control the output of the Sun or the shape of Earth's orbit. But we can control the amount of greenhouse gas in our atmosphere.

The Second Letter of Peter reads, "But the day of the Lord will come like a thief, and then the heavens will pass away with a loud noise, and the elements will be dissolved with fire,

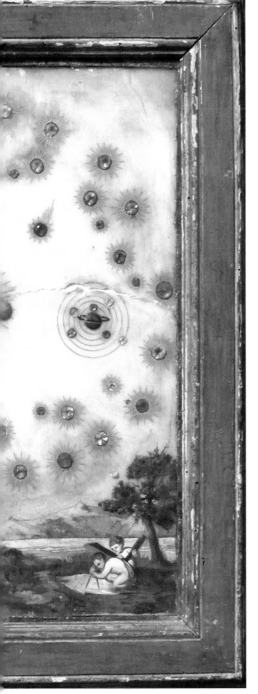

Left: This image of the solar system, dating probably from the 18th century, resides at the Vatican Observatory in Castel Gandolfo. It shows the planets as jewels embedded in a marble plaque.

and the earth and everything that is done on it will be disclosed" (2 Peter 3:10). What does science have to say about this vision? Is it true that there is a possibility the Earth could be destroyed by a gigantic meteorite?

First, remember that it is always problematic to try to treat a piece of Holy Scripture, out of context, as if it were a science book. Note that the verses immediately following the one quoted continue, "Since all these things are to be dissolved in this way, what sort of persons ought you to be in leading lives of holiness and godliness, waiting for and hastening the coming of the day of God, because of which the heavens will be set ablaze and dissolved, and the elements will melt with fire?" (2 Peter 3:11-12). In other words, the intent of the author is to teach us not about giant meteorites, but about how to live our lives.

That said, what does science tell us about the end of the Earth? We know for certain that our planet has a finite lifetime. (To quote Scripture again: "Heaven and earth will pass away, but My words will not pass away." — Matthew 24:35).

From observing other stars, we understand that a star like our Sun can only shine for about 10 billion years (see the images on page 29), and we are already halfway through that lifetime. Likewise, from observing the surfaces of other planets (where evidence for impacts is not worn away by wind and

water) we know that space debris like meteorites and comets do hit planets all the time. As Fr. Kikwaya describes in his chapter (page 156), even today we are being rained upon by cosmic dust, and larger rocks (see page 148) fall occasionally as well, to be collected and studied in our labs as meteorites.

Is it possible that Earth could be hit by an object big enough to wipe out most of life on the surface of this planet? Not only is it possible, it is certain. We know it has happened in the past, most famously with the extinction of the dinosaurs 65 million years ago. It is not a likely event — the odds are one in a hundred million of it happening in any given year — but it is not impossible. It would not take a particularly large asteroid to cause such damage. A body only a few tens of kilometers across could obliterate any place where it hit, and stir up enough dust to cut off sunlight over the whole Earth,

thus freezing the rest of us to death.

Only in the past 25 years have astronomers begun systematically to chart the motions of small bodies in the Solar System to see if we are at any immediate risk. There are easily a hundred thousand asteroids in our Solar System big enough to cause serious damage if they hit us, but virtually all

of them are in orbits that never come close to the Earth. The biggest risk may well come from comets, orbiting so far from us most of the time that we don't even know of their existence. We probably wouldn't see such a killer comet until it would be too late to do anything about it.

Top: Theodolite donated to the Vatican Observatory in 1891 by Fr. Giuseppe Buti; made by Troughton and Simms, London.
Above right: One of two analytical balances manufactured by Starke & Kammerer AG of Vienna given to Pope Pius IX in September 1934 by Johann Mörzinger, editor of the Kleinen Kirchenblattes, *the Viennese Catholic youth newspaper.*

However, if you are worried about dying from a comet impact, there are two things you should do: quit smoking, and wear your seat belt. Smoking and car accidents are far more likely to kill you than any rogue comet!

Many people have tried to identify the star of the Magi, spoken of in the second chapter of Matthew's Gospel, with comet Halley (last seen from Earth in 1986). Is this a true identification?

We don't know what the star of the Magi was, though many people have attempted to identify it with a particular comet, or a supernova, or a conjunction of planets. Halley's comet itself would have appeared in the year 12 B.C., which is probably too early to serve as an explanation for the star. Furthermore, comets were usually seen as unfavorable signs by the ancients, not consistent with something signifying the birth of a King.

For all we know, the description in Matthew's Gospel — the only source we have for the Magi — might be entirely symbolic. However, as Pope Benedict recently pointed out in a homily on Epiphany (see page 215) such symbols are important not in themselves but for the truths that they stand for. It is the birth of the Christ Child, not the star, that matters.

Why do we still need to use telescopes on Earth, like those of the Vatican Observatory in Castel Gandolfo or on Mt. Graham (near Tucson, Arizona), when we have more powerful telescopes now in orbit above the Earth, like the Hubble?

To perhaps oversimplify: discoveries are made on small telescopes, then confirmed and expanded by observing the same objects with bigger telescopes and space telescopes.

A discovery is, by definition, finding something new. If you already know what you're looking for, then it is not quite the same thing as casting a wide net in an area where you merely suspect there might be something interesting. The small telescopes act as the "wide net," while the big telescopes and the space telescopes can follow up on new discoveries to see more closely what the small telescopes have found.

Space telescopes, and the largest Earth-based telescopes like the 10-meter Keck Observatory in Hawaii or the Very Large Telescope (VLT) in Chile, have a long line of astronomers waiting to use them. Often you are fortunate just to get a few hours of time on the telescope before you have to turn it over to the next person in line. In addition, these telescopes are also designed to look very closely at specific, small areas of the sky.

By contrast, small telescopes like the VATT play two essential roles in astronomy. The first is to spend a lot of time surveying the sky. A survey project usually observes a large swath

....

Top: A solar and astronomical quadrant made by Abbé Berthiaud, c. 1880; donated to Pope Leo XIII in 1888. This device allows the calculation of the mean solar time, times of sunrise and sunset, and the location of the constellations at any time and place.

of the sky, and it can last for years; this is precisely what a space telescope cannot hope to do. The second is to "take a chance" and look at specific objects even when we are not sure if anything interesting is to be found there.

Often both roles are played at the same time. For example, the VATT was used in the early 2000s to make a survey of the broad colors of the brighter Trans-Neptunian Objects. To our surprise, we discovered that these objects seemed to be grouped into two distinct populations: red objects and gray objects. We (and our colleagues) were then able to get time on the Keck Telescope in Hawaii and the VLT in Chile to observe more of these bodies, including fainter ones, and we learned that the color populations were more complicated than we first thought. We used the 6.5-meter MMT telescope in Arizona to measure the specific spectral absorption lines of a few of these objects, and found evidence for methane ice on some of them but not on others. And we used the Spitzer Infrared Space Telescope to measure their temperatures and thus how dark they were, looking for a relationship between brightness and color. All of these observations turned a general pattern, red and gray objects, into a specific story of which objects show the strongest color differences and how that is related to where they orbit and how those orbits have changed with time. We needed the big telescopes to fill out the story; but we would never have known there was a story to be learned, without the original work on the VATT. ●

ABOUT THE VATICAN OBSERVATORY

Why is the Vatican interested in astronomy?

Initially it was for a practical reason, to reform the Julian Calendar, like so many national observatories that were started — e.g., to improve navigation at sea; later, at the establishment of the present form of the Vatican Observatory, in 1891, for an apologetic purpose, in the sense of defending the Catholic Church's positive regard for science; now, to join in doing good science in a way that is economically possible, given the Vatican's other concerns, as part of the consequence that the Incarnation of Christ applies to all human activity.

By the 1890s, a change had happened in the way science was done throughout the world. Science used to be the work of noblemen, doctors, and clergymen; who else had the free time and education to dedicate to studying nature? Indeed, the mundane work of a scientist — gathering and sorting data — is still called "clerical" work to this day. It was work done by clerics: by the clergy. But in the 19th century, as science became more and more of a technical (hence secular) job, a belief grew among many people that science and religion might be opposed. To counter that trend, Pope Leo XIII decided to establish a scientific institute that would show the world that the Church is not opposed to science, but, in fact, embraces and supports it.

Above: This painting by Donato Creti in 1711, which hangs in the Vatican Pinacoteca, contains the first color illustration of the Great Red Spot on Jupiter. It was a part of a series of paintings that secured Vatican support for an astronomical observatory in Bologna.
Right: The Vatican Advanced Technology Telescope (VATT) on Mt. Graham, Arizona.

Since not all science could be supported at once, the Pope chose a couple of astronomers to carry on this work. Why astronomers? Going back to the ancient and medieval universities, you find that astronomy was one of the subjects a student was expected to know before going on to learn theol-

226

ogy and philosophy. A long history exists of astronomy, the study of the universe, as a wonderful connection point between the philosophical and scientific yearnings to understand who we are and where we come from.

What is the mission statement for the Vatican Observatory?

From Leo XIII's letter *Motu Proprio* establishing the Vatican Observatory in 1891, his intent was to show that "the Church and her Pastors are not opposed to true and solid science, whether human or divine, but that they embrace it, encourage it, and promote it with the fullest possible dedication." In other words, it was to counteract claims of obscurantism on the part of the Church.

Nowadays, its mission is simply to do good science, for its own sake and within the worldwide community of scientists, and thus to be a bridge between Church and science. Where in the 19th century, the Church felt it had to tell scientists that they shouldn't be afraid of religion; now, our mission is to remind churchgoers that there's nothing to be afraid of in science. Too often a choice of science or religion has been given to the people in the pews. People know their religion and are comfortable with it, so that's their choice. As a result, some people believe they are being forced away from science, which is a tragedy since studying creation is a marvelous way of getting to know the Creator.

Pope John XXIII once said that our mission should be that of explaining the Church to astronomers and astronomy to the Church. We are like a bridge, a small bridge, between the world of science and the Church. As Pope Benedict XVI recommended to the Jesuits on the occasion of their most recent general congregation, we should be "men on the cutting edge." Thus,

the observatory has this mission: to be on the frontier between the world of science and the world of faith, to give testimony that it is possible to believe in God and to be good scientists.

Is the Church looking for something in space: aliens to baptize, or a "sign from above," so to speak?

No — despite some people's suspicions!

Does your astronomy ever come in conflict with your religion? Don't you have to believe in what the Bible says about how the universe was made?

As we show in several chapters of this book (see especially "The Popes and Astronomy"), the whole of Catholic tradition has been to reject this kind of biblical "literalism."

The Bible is not a science textbook. When it was written, the idea of "science textbooks" didn't even exist. And

....

Top: Astrolabe, of the Flemish school, undated but probably second half of the 16th century; given to Pope Leo XIII in 1888.

to confuse it with a science book does the Bible no honor. Great works of literature, philosophy, or theology are studied as intently today as when they were written, but no scientist actually learns science today from reading, say, Newton's original works. Science books go out of date only a few years after they are written and must be constantly revised. By contrast, the Bible is timeless. Thus, you can see that the Bible is fundamentally different from a science book.

We do take the Bible seriously. It teaches us that the physical universe was made by God, in an orderly fashion, Who found that His creation was good, and Who indeed so loved this world that He sent His only Son. This motivates us to study the physical universe, in order to become closer to its Creator. The Bible tells us Who made the universe; science tells us how He did it.

Would the discovery of extraterrestrial intelligence — aliens — affect your faith?

The first and most important fact we have to confront in the whole question of "extraterrestrial intelligence" is this: we don't know. Of all the planets we've found orbiting other stars, it's not clear if any of them are suitable places for life as we know it. On none of them, nor indeed anywhere closer to us in our own Sun's system of planets, have we ever found evidence that completely, incontrovertibly proves life originated in some place other than just here on Earth. As far as we know for sure, we could be alone.

Fr. Ernan McMullin, a philosophy professor at Notre Dame with a background in physics, has discussed the possible impact on Christian theology of discovering extraterrestrials, and he concludes only that it would certainly inspire theologians to develop

Why does it seem that it's been the Jesuits who have been most involved with the sciences for the Church?

Compared with other orders, Jesuits have more men, more flexibility in the type of work undertaken, and a tradition of scholarship. All of those reasons contributed to the decision of the Papacy to entrust the Vatican Observatory to the Jesuit order in 1904. However, note that Barnabites, Oratorians, and Augustinians all were involved in the Vatican Observatory in its early days. And, even today, one of the staff of the observatory is a priest from the Diocese of Padua and is not a Jesuit.

How do you report your findings? Do you communicate them directly to the Holy See?

new ways of thinking about topics like original sin, the immortality of the soul, and the meaning of Christ's redemptive act. But, as he points out, there is already a voluminous literature, and hardly a consensus, on these points among theologians even today, without ETs!

Giuseppe Tanzella-Nitti, an astronomer and Opus Dei priest who teaches theology at the Pontifical University of the Holy Cross in Rome, comes to the same conclusion. He has written a lengthy entry on extraterrestrial life in the *Dizionario Interdisciplinare di Scienza e Fede* (*The Interdisciplinary Dictionary of Science and Faith*, of which he was an editor). But at the end, he concludes by saying (in my translation of his Italian), "the last word on the question of extraterrestrial life will not come from theology, but science. The existence of intelligent life on planets other than the Earth neither rules in, nor rules out, any theological principle. Theologians, like the rest of the human race, will just have to wait and see."

The mere possibility of intelligent life elsewhere puts a human (or at least, human-like) face on the far better established astronomical observation of the enormity of our universe. For us Catholics, the thoughts that come from contemplating this question, in the absence of any firm answers, should lead us to focus on realizing God's greatness and His special love for each of us.

Above: The observatory met at a retreat house in Calascio, Italy, in 2008, to plan its future activities. Here Fr. Giuseppe Koch (left), rector of the community, chats with Fr. José Funes, director of the Vatican Observatory.

Our scientific work is published in the regular international journals of astronomy, refereed by our fellow astronomers. At the end of each year, our accomplishments for the year are written up (in Italian) in an annual report and presented to the Holy See. (An expanded version of this annual report, in English, is available on our website: http://vaticanobservatory.org)

Has an astronomer ever become Pope?

Certainly several Popes have had an interest in astronomy, as we have noted in an earlier chapter. Pope Pius X, for example, had studied mathematics and astronomy, and enjoyed making sundials; and Pope Pius XII was an enthusiastic amateur astronomer.

However, one could well describe a much earlier Pope as one of the leading astronomers and scientists of his day. Pope Sylvester II's scholarship was well ahead of his time, to such an extent that in popular lore he was thought to be a wizard and magician! Born Gerbert d'Aurillac in France in the middle of the 10th century, as a young man he entered the monastery of St. Gerald of Aurillac in Spain, where he first came in contact with Arabic learning. He was especially fascinated by the Arabic scholars of mathematics and astronomy in Cordoba, then under the rule of the Arabs. He later taught astronomy at the Cathedral School of Rheims, where he also served briefly as bishop. Later he was a tutor to Holy Roman Emperor Otto III, and served as archbishop of Ravenna, before being elected Pope in 999; he served until his death in 1003.

He is probably most famous for introducing to Europe the use of Arabic numbers and the abacus. In the realm of astronomy, he introduced to Europeans the use of the armillary sphere, an early device used to demonstrate the positions of the Sun and planets during the year.

How expensive is it for the Vatican to support an astronomical observatory? Where does your funding come from?

The Vatican Observatory is a branch of the formal government of the Vatican City State. Its budget makes up about one half of one percent of the total annual budget of the state — about the same proportion as the NASA budget in the U.S. government. However, since the members of the observatory are Jesuit priests and brothers living in community under a vow of poverty (the daily stipend per astronomer to cover regular living expenses like food and shelter is less

than 20 euros per day), this money goes a lot further and supports a lot more research than would be possible at a traditional observatory.

The Vatican Observatory Advanced Technology Telescope is funded separately from the rest of the observatory. The original funds to build the telescope came from private donors, and the annual budget to run the telescope comes from the Vatican Observatory Foundation, which maintains an endowment supported by private donors.

Is the Vatican Observatory recognized and accepted by the international astronomical community?

Yes. Since its inception, the observatory has worked in close collaboration with other astronomers around the world. One of its first, and largest, projects was participation in the international program organized by the Paris Observatory to make a photographic map of the sky, known as the Cart du Ciel. Every major European observatory that volunteered to participate was assigned a region of the sky to photograph; the Vatican Observatory was welcomed,

Above: Modern heirs of Boscovich (see p. 138): Vatican astronomers Fr. Richard Boyle, S.J., and Fr. Jean-Baptiste Kikwaya, S.J.
Top: Filar micrometer, used to measure relative positions of stars seen in a telescope, made by Merz & Sons, Munich, in 1890 and acquired by the Vatican Observatory, c. 1900.

even though many states at that time did not yet recognize the Holy See as a nation independent of Italy. The well-respected international journal of spectrochemistry, *Spectrochimica Acta*, was established at the Vatican Observatory in the late 1930s and printed by the Vatican in the years following World

War II, when facilities for producing scientific journals were affected by the scarcities after the war.

Vatican Observatory astronomers, like astronomers everywhere, work in collaboration with colleagues from outside their home institution. Today, our collaborators come from countries all over the world, including the United Kingdom, France, Finland, Lithuania, Poland, South Africa, Brazil, Argentina, Chile, and, of course, Italy and the United States. And the observatory has sponsored international meetings in both Rome and Tucson on subjects like cosmology, galactic evolution, stellar classification, and meteoritics.

One sign of the degree of respect that the international astronomical community has for the observatory is that throughout its history, members of the observatory have been elected by their peers to offices within a number of astronomical societies. Organizations where Vatican astronomers have held office include the American Astronomical Society, the Division for Planetary Sciences, the Meteoritical Society, and the International Astronomical Union.

Is one of the roles that the Vatican Observatory plays in astronomy that of verifying what other astronomers are telling the public in order to make sure what they are reporting to the public is actually true?

No, not formally. The watchdog for astronomers is the astronomical community itself, and so by being part of that community the Vatican Observatory astronomers join in that watchdog role. For example, we join in refereeing papers by other astronomers before they are published in the main journals, just as our papers are so refereed; and we write book reviews and critiques along with the rest. And because we are not competing with other scientists for funding from their national agencies, we are often asked to review funding proposals and serve on review panels to determine how monies from American funding agencies such as the National Science Foundation and NASA, and various European space agencies, should be allocated.

How did you personally become interested in astronomy, and what led you to become a Jesuit astronomer?

The stories are as varied as the different members of the observatory staff. Here are our stories:

....

Top: A heliophanograph, a device to record the hours of sunshine in a day: the crystal ball focuses sunlight onto a paper strip, leaving a burn mark, while cloudy periods leave no mark. Manufactured by J. Hicks, London, in the latter half of the 19th century, and given to the Vatican Observatory before 1897.

Chris Corbally: "Through an interest in science at my high school, though the actual observatory was in disrepair at that time. I entered the Jesuits upon leaving school, knowing that they had a tradition of combining scientific work and priestly service. With superiors' consent, I gradually focused on the possibility of astronomy and particularly the Vatican Observatory during the course of my further studies."

Guy Consolmagno: "In college, I thought a lot about both options, and after long, serious prayer I realized that God's first call to me was to be a scientist, not a priest. Thus, I changed schools (and majors), moving to MIT to study planetary sciences, and then to Arizona for a doctorate. After 15 years in the field, teaching in many places, I finally felt ready to enter a religious order — but as a brother, not a priest. I made this decision at a time in my life when I was very happy teaching at a wonderful little school, secure in my life and my occupation and ready to take the next step to use my talents in the direction I felt God calling me." ●

ACKNOWLEDGMENTS

This book combines the creative efforts of many people. First and foremost I must thank the authors, my fellow astronomers at the Vatican Observatory, and my brothers in religious life. And as we have all noted, our work is done in collaboration with fellow astronomers around the world, too many to mention here, but all of whom have contributed to our work.

We acknowledge here, as we also credit throughout the book, the institutions and professional photographers whose images have helped us tell our story. Notable are, of course, the various Papal photographers through the years.

I would like to specially acknowledge here our friends and visitors, talented amateur photographers, who have so generously shared their photographs. These include Alex Lovell-Troy, Jim Scotti, Alfredo Matacotta Cordella, Sassone Corsi, Roelof de Jong, and David Wang. Their work has added so much beauty to this book.

Ron Dantowitz, an extraordinary astrophotographer at the Clay Center Observatory in Brookline, Massachusetts, provided not only some wonderful images of the Sun but also beautiful shots of the telescopes and their environs in Castel Gandolfo.

Many of the images from the VATT were taken by Matt Nelson, now at the University of Virginia, who worked for many years at the VATT and who first showed us the wonderful qualities of

our telescope as an imager of celestial wonders. Some of his images were taken with the help of docents from the Discovery Park Museum in nearby Safford, Arizona.

Tijl Kindt, a student of the 2007 Vatican Observatory Summer School, graciously allowed us to use his images of that summer school. (I also thank the summer school students, who gave us their permission to use their pictures here.) In addition, many of the pictures of Castel Gandolfo are also his.

We thank Lori Styles and the press office of the University of Arizona, who provided several of the photos of the VATT under construction.

The beautiful design of this book is the work of Giancarlo Olcuire, with whom I feel honored to have been able to work.

And finally, as we have already noted, the idea for this book came from Giovanni Cardinal Lajolo, whose continued support has made it all possible.

Br. Guy Consolmagno, S.J.
January 2009,
The International Year of Astronomy